Recent Titles in This Series

(Continued in the back of this publication)

MEMOIRS
of the
American Mathematical Society

Number 534

Orthogonal Decompositions
and Functional Limit Theorems
for Random Graph Statistics

Svante Janson

September 1994 • Volume 111 • Number 534 (third of 5 numbers) • ISSN 0065-9266

American Mathematical Society
Providence, Rhode Island

1991 *Mathematics Subject Classification.*
Primary 05C80; Secondary 60F05.

Library of Congress Cataloging-in-Publication Data

Janson, Svante.
Orthogonal decompositions and functional limit theorems for random graph statistics / Svante Janson.
   p. cm. – (Memoirs of the American Mathematical Society, ISSN 0065-9266; no. 534)
   "Volume 111, number 534 (third of 5 numbers)."
   Includes bibliographical references.
   ISBN 0-8218-2595-X
   1. Random graphs. 2. Central limit theorem. I. Title. II. Series.
QA3.A57 no. 534
[QA166.17]
510 s–dc20   94-17088
[511′.5]   CIP

Memoirs of the American Mathematical Society

This journal is devoted entirely to research in pure and applied mathematics.

Subscription information. The 1994 subscription begins with Number 512 and consists of six mailings, each containing one or more numbers. Subscription prices for 1994 are $353 list, $282 institutional member. A late charge of 10% of the subscription price will be imposed on orders received from nonmembers after January 1 of the subscription year. Subscribers outside the United States and India must pay a postage surcharge of $25; subscribers in India must pay a postage surcharge of $43. Expedited delivery to destinations in North America $30; elsewhere $92. Each number may be ordered separately; *please specify number* when ordering an individual number. For prices and titles of recently released numbers, see the New Publications sections of the *Notices of the American Mathematical Society*.

Back number information. For back issues see the *AMS Catalog of Publications*.

Subscriptions and orders should be addressed to the American Mathematical Society, P. O. Box 5904, Boston, MA 02206-5904. *All orders must be accompanied by payment.* Other correspondence should be addressed to Box 6248, Providence, RI 02940-6248.

Memoirs of the American Mathematical Society is published bimonthly (each volume consisting usually of more than one number) by the American Mathematical Society at 201 Charles Street, Providence, RI 02904-2213. Second-class postage paid at Providence, Rhode Island. Postmaster: Send address changes to Memoirs, American Mathematical Society, P. O. Box 6248, Providence, RI 02940-6248.

10 9 8 7 6 5 4 3 2 1   99 98 97 96 95 94

CONTENTS

ABSTRACT

We define an orthogonal basis in the space of real-valued functions of a random graph, and prove a functional limit theorem for this basis. Limit theorems for other functions then follows by decomposition. The results include limit theorems for the two random graph models $G_{n,p}$ and $G_{n,m}$, as well as functional limit theorems for the evolution of a random graph and results on the maximum of a function during the evolution. Both normal and non-normal limits are obtained. As examples, applications are given to subgraph counts and to vertex degrees.

Partially supported by the Göran Gustafsson Foundation for Research in Natural Sciences and Medicine.

1991 *Mathematics Subject Classification.* 05C80; 60F05.

Key words and phrases. Random graphs; subgraph counts; vertex degrees; functional limit laws; martingale limit theorems.

I. FOUNDATIONS

1. Introduction

We consider a random graph which evolves in time as follows. Assume that n is a positive integer. Let K_n be the complete graph with vertices $\{1, \ldots, n\}$ and let T_e, where e ranges over the $\binom{n}{2}$ edges in K_n, be independent random variables that are uniformly distributed on $(0, 1)$. We interpret T_e as the time the edge e appears and let $G_n(t)$ be the (random) graph consisting of the vertices $\{1, \ldots, n\}$ and the edges in K_n that have appeared at time t. $G_n(t)$ is our random graph process, and we will use the following notations, for $t \geq 0$ and e an edge in K_n (which we denote by $e \in K_n$),

$$I_e(t) = I(e \in G_n(t)) = I(T_e \leq t), \tag{1.1}$$

$$I'_e(t) = I_e(t) - \mathrm{E}\, I_e(t) = I_e(t) - t, \qquad 0 \leq t \leq 1. \tag{1.2}$$

(It will be convenient to allow also $t > 1$; we let $I_e(t) = 1$ and $I'_e(t) = 0$ for $t > 1$.) One advantage of this model is that it contains both standard types $G_{n,p}$ and $G_{n,m}$ of random graphs. (See e.g. Bollobás [5] for the definitions and other properties of these random graphs.) In fact, if we fix $t \in [0, 1]$ then $G_n(t)$ is a Bernoulli random graph $G_{n,t}$ where all edges appear independently with equal probabilities t. On the other hand, if $T_{(m)}$ is the mth order statistic of $(T_e)_{e \in K_n}$, i.e. the time the mth edge appears, then $G_n(T_{(m)})$ is a random graph $G_{n,m}$ with m edges distributed at random. (Note that T_e a.s. are distinct.) Thus, as we shall see in detail later, results for both $G_{n,p}$ and $G_{n,m}$ can be derived from results for $G_n(t)$. Note also that the sequence $\{G_n(T_{(m)})\}_{m=0}^{\binom{n}{2}}$ yields the random graph process introduced by Erdős and Rényi [6], where edges are added successively at random.

REMARK 1.1. The distribution of T_e determines the time scale of the process but is otherwise irrelevant, since different choices are equivalent by a (deterministic) change of time. We find it convenient to fix the distribution and will in this paper only consider $T_e \sim U(0, 1)$. Note, however, that other choices may at times be convenient; it is also possible to let the distribution depend on n. For example, in Theorem 1, we rescale the time with a factor p_n depending on n. An (obviously equivalent) alternative would be to let T_e be uniform on $[0, 1/p_n]$ and then study the resulting processes (for fixed t). Another attractive choice is to let T_e be exponentially distributed, which makes the random graph process a homogeneous Markov process; this version was introduced by Stepanov [25].

Received by the editor June 21, 1992.

1

We will study the random graph process through the following family of statistics. Let H be a graph and denote its number of vertices by $v(H)$ and its number of edges by $e(H)$. Consider the $(n)_{v(H)}$ different injective mappings from the vertices of H into $\{1, \ldots, n\}$. Each such mapping φ maps H onto a subgraph $\varphi(H)$ of K_n, and we define

$$S_n(H; t) = \sum_{\varphi} \prod_{e \in \varphi(H)} I'_e(t). \tag{1.3}$$

In other words, we sum $\prod_{e \in H_1} I'_e$ over all copies H_1 of H in K_n, counted with multiplicities $\operatorname{aut}(H)$ (the number of automorphisms of H). Note that if we replace I'_e by I_e in (1.3), we obtain the number of copies of H in $G_n(t)$ (counted with multiplicities). The variables defined in (1.3) are, however, much more useful for our purposes, mainly because they are orthogonal (see below); in fact, they may be regarded as an orthogonalized version of the subgraph counts.

It is obvious that $S_n(H; t)$ depends only on the isomorphism type of H. Hence we may, and will, regard $\{S_n(H; t)\}$ as a family of stochastic processes (i.e. random functions of t) which are indexed by *unlabelled* graphs H. The simplest examples are

$$S_n(\emptyset; t) = 1 \qquad \text{(trivial but useful)} \tag{1.4}$$

$$S_n(K_1; t) = n \qquad \text{(cf. Proposition 1.1(i))} \tag{1.5}$$

$$S_n(K_2; t) = 2 \sum_{e \in K_n} I'_e(t) = 2\big(e(G_n(t)) - \tbinom{n}{2}t\big), \qquad 0 \le t \le 1. \tag{1.6}$$

Since the variables $I'_e(t)$ are independent and have mean 0, two products $\prod I'_e(t)$ are orthogonal unless they coincide, and the following results follow easily. We let $(n)_v = n(n-1)\cdots(n-v+1)$.

PROPOSITION 1.1.

(i) If H has v_0 isolated vertices, and $\overline{H}$ is the graph obtained by removing them, then

$$S_n(H; t) = (n - v(\overline{H}))_{v_0} S_n(\overline{H}; t). \tag{1.7}$$

(ii) If H has at least one edge, then

$$\mathrm{E}\, S_n(H; t) = 0. \tag{1.8}$$

(iii) If $H \ne \emptyset$ lacks isolated vertices, then

$$\operatorname{Var} S_n(H; t) = \mathrm{E}\, S_n(H; t)^2 = \operatorname{aut}(H)(n)_{v(H)}(t(1 - t))^{e(H)}, \qquad 0 \le t \le 1. \tag{1.9}$$

(iv) If H and K are different unlabelled graphs, and both H and K lack isolated vertices, then $S_n(H; t)$ and $S_n(K; t)$ are orthogonal:

$$\operatorname{Cov}(S_n(H; t), S_n(K; t)) = \mathrm{E}\, S_n(H; t) S_n(K; t) = 0. \tag{1.10}$$

$\square$

It is easy to extend (1.9) and (1.10) to $\mathrm{Cov}(S_n(H;t), S_n(H;u))$ for two different times t and u; if H and K are non-empty and lack isolated vertices,

$$\mathrm{Cov}(S_n(H;t), S_n(K;u)) = \delta_{HK} \, \mathrm{aut}(H)(n)_{v(H)} \big(t(1-u)\big)^{e(H)}, \qquad 0 \le t \le u \le 1. \tag{1.11}$$

Our fundamental result, Theorem 1 (stated and proved in Section 3), is a functional limit theorem which shows that, with appropriate conditions and normalization, the processes $S_n(H;t)$ converge in distribution to some limit processes. Using this limit theorem, we then may obtain limit theorems for many other graph statistics such as subgraph counts. The general procedure is as follows; see Sections 4 and 5 for details. Let ψ be any graph statistic, i.e. a function assigning to each graph a real number that depends only on the isomorphism type of the graph. Then the random variable $\psi(G_n(t))$ may be written as a linear combination of $S_n(H;t)$ for some graphs H,

$$\psi(G_n(t)) = \sum_H \hat{\psi}_n(H;t) S_n(H;t), \tag{1.12}$$

for some coefficients $\hat{\psi}_n(H;t)$ depending on n and t. In many cases it is now easy to use the orthogonal decomposition (1.12) and Theorem 1 below to obtain the asymptotic distribution of $\psi(G_n(t_n))$ as $n \to \infty$, where $\{t_n\}$ is a given sequence.

The simplest case is when only a finite set of graphs H, independent of n, is needed in the decomposition (1.12). The asymptotic behaviour of $\psi(G_n(t_n))$ then follows from Theorem 1 and a knowledge of the asymptotic behaviour of the coefficients $\hat{\psi}_n(H;t)$. Both normal and non-normal limits may be obtained by this procedure. In fact, $\psi(G_n(t_n))$ is asymptotically normal if and only if the terms with H connected dominate the decomposition (1.12).

The method extends (through a truncation argument) to more general situations, see Section 5. Note that in order to apply it to a given ψ, it is not necessary to compute all coefficients $\hat{\psi}(H;t)$. Usually it suffices to find the order of magnitude of the most important term(s), and to show that the others are small.

Furthermore, we will also obtain limit results for $G_{n,m}$ (Section 6) and the whole process $G_n(t)$ (Sections 8 and 9). Some results on moment convergence are given in Section 7. The proofs of some of the theorems (in particular the main one) uses methods and results from the theory of stochastic processes and may appear difficult to non-specialists. (The basic definitions and results that are used are collected in Section 2, which may be skipped by those not interested in the technicalities.) We hope, nevertheless, that the theorems will be easy to apply; the aim has been to use only easily verified assumptions such as variance estimates. In the same spirit, the theorems on convergence for $G_{n,m}$ are stated using conditions only for $G_{n,p}$, where calculations generally are easier.

Some illustrative applications are given in the final sections, where we give new proofs of some old results and add a number of new results. In Section 10, we consider the number of induced subgraphs of a given type, and prove asymptotic normality both for $G_{n,p}$ and $G_{n,m}$. We consider also induced subgraphs and obtain similar results, with some striking exceptions which are readily explained by our method. In Section 11 we consider the number of vertices of a given degree and, again, obtain asymptotic normality for $G_{n,p}$ and $G_{n,m}$. Further results include joint distributions and asymptotic normality of the maximum during the evolution. Some further examples are briefly treated in Section 12.

We consider in this paper only simple, undirected graphs. The methods apply, however, also to other models of random graphs, yielding similar results for e.g. bipartite graphs, directed graphs or multigraphs. We leave such extensions to the reader. See [15] for results of this type proved by the method of moments, and [12] for some examples using the method presented here.

All unspecified limits in this paper are taken as $n \to \infty$. Note that all results hold also if we consider only a fixed subsequence (n_k).

Acknowledgements. I am indebted to Andrzej Ruciński for his pertinent questions which have led to considerable improvement of the results in this paper. I also thank Torgny Lindvall for helpful remarks.

2. Preliminaries

Graphs. All graphs considered in this paper are finite, simple and undirected. If G is a graph, we denote its vertex set by $V(G)$ and its edge set by $E(G)$; we let $v(G)$ be the number of vertices (the order of G), $e(G)$ the number of edges, and $\mathrm{aut}(G)$ the number of automorphisms. We include the empty graph $\emptyset$, with $v(\emptyset) = e(\emptyset) = 0$. We further define $v_0(G)$ to be the number of isolated vertices in G, and

$$m(G) = \max\{\tfrac{e(H)}{v(H)} : \emptyset \neq H \subseteq G\}. \tag{2.1}$$

(For completeness, we define $m(\emptyset) = 0$.)

A graph G is *discrete* if it has no edges, i.e. $e(G) = 0$ or, equivalently, $v_0(G) = v(G)$.

We use *unlabelled* graphs to index e.g. the terms in the orthogonal decomposition (1.12). Unlabelled graphs may be defined as equivalence classes of labelled graphs: thus two different unlabelled graphs are non-isomorphic. Let $\mathcal{U}$ denote the set of all unlabelled graphs; since there are only a finite number of unlabelled graphs of each order, $\mathcal{U}$ is a countable set. We will mainly use the subset $\mathcal{U}^0 = \{H \in \mathcal{U} : v_0(H) = 0\}$ of unlabelled graphs without isolated vertices. $\mathcal{U}^0$ contains one graph of order 0, viz. $\emptyset$, no graph of order 1, one graph of order 2 (K_2), two graphs of order 3 (K_3 and P_2), etc. We let $\mathcal{U}_k^0$ denote the set of graphs in $\mathcal{U}^0$ with k vertices, and let $\mathcal{U}_k^c$ denote the subset of connected graphs with k vertices.

The disjoint union of two unlabelled graphs G and H will be written $G + H$. We will also write kH for the disjoint union of k copies of H.

A *graph statistic* is a (real-valued) function $\psi(G)$ of graphs that depends only on the isomorphism type of G. Equivalently, it may be regarded as a function on the set of unlabelled graphs.

REMARK 2.1. Graphs play several different roles in this paper, and these must not be confused. The basic objects of study are the (labelled) random graphs $G_n(t)$, $G_{n,p}$ and $G_{n,m}$; the processes $S_n(H;t)$ are indexed by (unlabelled) graphs H; in Section 10 we treat the number of subgraphs of a random graph that are isomorphic to a fixed (unlabelled) graph G. (K_n will be used both for a labelled graph, as in the introduction, and an unlabelled graph. This should not cause any problems.)

The Skorokhod topology. Let I be any interval (open, closed or semi-open; finite or infinite) contained in the extended real line $[-\infty, \infty]$ and let $D(I)$ be the set of all (real-valued) functions on I that are right-continuous and have left-hand limits. (We are mainly interested in the cases $[0, 1]$, $[0, \infty)$, $[0, \infty]$.) The standard topology on D is the Skorokhod topology. A description of the topology is that $f_n \to f$ in D if and only if there exist strictly increasing continuous mappings λ_n of I onto itself such that $\lambda_n \to \iota$ (the identity mapping) uniformly on I and $f_n \circ \lambda_n \to f$ uniformly on compact subsets of I. This topology makes D into a Polish space, i.e., it is possible to describe the topology with a complete, separable metric. This is technically important, but the metric is complicated and seldom used explicitly. When the limit function is continuous (the only case of real importance to us), there is a simpler criterion: if f is continuous then $f_n \to f$ in D if and only if $f_n \to f$ uniformly on compact sets. For these and other properties of the Skorokhod topology, see e.g. [4], [7], [8], [20].

The Borel σ-field in $D(I)$ concides with the σ-field generated by the point evaluations $f \to f(t)$, $t \in I$. Hence a mapping $\Omega \to X$ of some measurable space into $D(I)$ is measurable if and only if the mapping $\Omega \to X(t)$ is measurable for every fixed $t \in I$.

REMARK 2.2. It would be much simpler if we instead could use the topology of uniform convergence on compact sets (which thus agrees with the Skorokhod topology for convergence to continuous functions). This is, however, not possible for technical reasons. The Borel σ-field would be too large and it is not even possible to define $I_e(t)$ as a random function in $D([0, 1])$ with the uniform topology, see [4, Section 18].

Let $X_n(t)$ and $X(t)$ be random functions in $D(I)$. Then $X_n(t) \xrightarrow{d} X(t)$ in $D(I)$ implies that $X_n(t_0) \xrightarrow{d} X(t_0)$ for every fixed $t_0 \in I$ such that $X(\cdot)$ is a.s. continuous at t_0 (but not necessarily otherwise). Furthermore, the joint distribution of $(X_n(t_1), \ldots, X_n(t_m))$ converges for every such $t_1, \ldots, t_m \in I$. We can also consider a sequence $t_0^{(n)} \to t_0$, or even let the times be random as in the following result.

PROPOSITION 2.1. Suppose that $X_n(t)$ and $X(t)$ are random functions in $D(I)$ and that τ_n are random elements of I. If $X_n(t) \xrightarrow{d} X(t)$ in $D(I)$ and $\tau_n \xrightarrow{p} t_0 \in I$, with $X(t)$ a.s. continuous at t_0, then $X_n(\tau) \xrightarrow{d} X(t_0)$. Furthermore, $X_n(\tau) - X_n(t_0) \xrightarrow{p} 0$. $\square$

Addition is not continuous in D, but $f_n \to f$ and $g_n \to g$ with f and g continuous implies $f_n + g_n \to f + g$. (In fact, it suffices that one of f and g is continuous or just that they have no common discontinuities.) The same is true for multiplication, which implies the following result on weak convergence.

PROPOSITION 2.2. Suppose that $X_n^i(t)$ and $X^i(t)$, $i = 1, \ldots, m$, are random functions in $D(I)$ such that every X^i is continuous and $X_n^i \xrightarrow{d} X^i$ as

$n \to \infty$, *jointly for* $i = 1, \ldots, m$. *Then* $q(X_n^1, \ldots, X_n^m) \xrightarrow{d} q(X^1, \ldots, X^m)$ *for every polynomial* q *in* m *variables, with joint convergence for any finite set of polynomials.* □

We will also use the following results, relating weak convergence in the Skorokhod topology on different intervals. These results are partly given in the references given above (see in particular [20]); they may all be proved by the same methods but we omit the details.

PROPOSITION 2.3.

(i) *If* $X_n \xrightarrow{d} X$ *in* $D(I)$ *and* X *is a.s. continuous, then* $X_n \xrightarrow{d} X$ *in* $D(J)$ *for every subinterval* $J \subset I$.

(ii) *If* $I = \cup J_m$ *for an increasing sequence of intervals* J_m, *and* X_n *and* X *are random elements of* $D(I)$ *such that* $X_n \xrightarrow{d} X$ *in* $D(J_m)$ *for every* J_m, *then* $X_n \xrightarrow{d} X$ *in* $D(I)$.

(iii) *If* $I = J_1 \cup J_2$ *where* J_1 *and* J_2 *are intervals such that* $J_1 \cap J_2 \neq \emptyset$, *and* X_n *and* X *are random elements of* $D(I)$ *such that* $X_n \xrightarrow{d} X$ *in* $D(J_1)$ *and* $D(J_2)$, *jointly, then* $X_n \xrightarrow{d} X$ *is* $D(I)$. □

(The joint convergence in (iii) means that the pair of restrictions $(X_n|_{J_1}, X_n|_{J_2})$ converges to $(X|_{J_1}, X|_{J_2})$ in $D(J_1) \times D(J_2)$.)

PROPOSITION 2.4. *If* X_n *and* X *are random functions in* $D([a, b])$, *and* X *is a.s. continuous on* $[a, b]$, *then* $X_n \xrightarrow{d} X$ *in* $D([a, b])$ *if and only if*

(i) $X_n \xrightarrow{d} X$ *in* $D([a, b))$, *and*

(ii) $\limsup_{n \to \infty} P(\sup_{u < t < b} |X_n(t) - X_n(b)| > \varepsilon) \to 0$ *as* $u \to b$ *for every* $\varepsilon > 0$.

Furthermore, if X *is defined on* $[a, b)$ *only, and* (i) *and* (ii) *hold, then* $X(b) = \lim_{t \to b-} X(t)$ *exists a.s., and* $X_n \xrightarrow{d} X$ *in* $D([a, b])$.

The same holds if the intervals $[a, b]$ *and* $[a, b)$ *are replaced by* $(a, b]$ *and* (a, b), *or, with the obvious modification of* (ii), *by* $[a, b]$ *and* $(a, b]$ *or* $[a, b)$ *and* (a, b). □

The preceding discussion extends without changes to the space $D(I, \mathbf{R}^k)$ of vector valued functions (or, more generally, the space $D(I, E)$ of functions with values in an arbitrary Polish space E). $D(I, \mathbf{R}^k)$ and $(D(I, \mathbf{R}))^k$ can be identified as sets, and their Borel σ-fields coincide, but their topologies differ. Thus a random function $X(t)$ of $D(I, \mathbf{R}^k)$ is the same as k random functions $X^{(1)}(t), \ldots, X^{(k)}(t)$ in $D(I, \mathbf{R})$. Furthermore, the two topologies coincide for convergence to continuous limits. Hence, if $X(t)$ is a.s. continuous, then $X_n(t) \xrightarrow{d} X(t)$ in $D(I, \mathbf{R}^k)$ if and only if $X_n^{(i)}(t) \xrightarrow{d} X^{(i)}(t)$ in $D(I, \mathbf{R})$, jointly for $i = 1, \ldots, k$.

Subsequences. We use several times the following fact. (In fact, this holds for convergence of sequences in any topological space, and has nothing to do with random variables.)

PROPOSITION 2.5. *A sequence X_n of random variables or functions converge in distribution to X if and only if every subsequence has a subsubsequence that converges to X.* □

Continuous time martingales. A *martingale* on an interval I is a stochastic process $X(t) \in D(I)$ such that $E|X(t)| < \infty$ for every $t \in I$ and $E(X(t) \mid \mathcal{F}_s) = X(s)$, $s \leq t$, for some increasing family $(\mathcal{F}_t)_{t \in I}$ of σ-fields (included in the originally given σ-field $\mathcal{F}$). It is customary to assume that every $\mathcal{F}_t$ contains all null sets in $\mathcal{F}$, and that the family is right-continuous, i.e., that $\cap_{s>t}\mathcal{F}_s = \mathcal{F}_t$; these are known as the 'usual conditions'. In this paper, $\mathcal{F}_t$ will always be the σ-field generated by all $I_e(s)$ with $s \leq t$ (completed with all null sets).

Doob's inequality says that if $M(t)$ is a martingale on $[a, b]$, and $1 < r < \infty$, then

$$\| \sup_{a \leq t \leq b} |M(t)| \|_r \leq \frac{r}{r-1} \|M(b)\|_r, \tag{2.2}$$

where $\|X\|_r = (E|X|^r)^{1/r}$. (For $r = 1$ there is a corresponding weak estimate.)

We will use the notion of the *quadratic variation* $[M, M]_t$ of a martingale and, more generally, the *quadratic covariation* $[M, N]_t$ of two martingales. For the general definition and discussion we refer to [23, Section II.6]; when M and N are processes of finite variation, as in our applications,

$$[M, N]_t = \sum_{0 \leq s \leq t} \triangle M(s) \triangle N(s), \tag{2.3}$$

where $\triangle M(s) = M(s) - M(s-)$ and $\triangle N(s) = N(s) - N(s-)$ are the jumps at s (we set $M(0-) = N(0-) = 0$). (The sum is at most countable.) In particular, for two processes of finite variation without common jumps, $[M, N]_t = 0$.

We define, for $0 \leq t < 1$,

$$\check{I}_e(t) = (1 - t)^{-1} I'_e(t) = \frac{I_e(t) - t}{1 - t} \tag{2.4}$$

and

$$\check{S}_n(H; t) = (1 - t)^{-e(H)} S_n(H; t) = \sum \prod_{e \in H_1} \check{I}_e(t) \tag{2.5}$$

where we sum over all copies H_1 of H in K_n, counted with multiplicites.

LEMMA 2.1. *$\check{I}_e$ and $\check{S}_n(H)$ are martingales on $[0, 1)$.*

Proof. It is easy to see that $\check{I}_e$ is a Markov process with $E(\check{I}_e(t) \mid \check{I}_e(s)) = \check{I}_e(s)$, $0 \leq s \leq 1$, which implies that $\check{I}_e$ is a martingale (with respect to $(\mathcal{F}_t)$). Consequently, $\check{S}_n(H)$, which is a sum of products of independent martingales, is also a martingale. □

Our results are based on a powerful martingale convergence theorem by Jacod and Shiryaev [8]. We state a special case which is convenient for our application.

PROPOSITION 2.6. *Let I be an interval $[0, b]$ or $[0, b)$, $0 < b \le \infty$. Assume that for each n, $M_n(t) = (M_n^i(t))_{i=1}^d$ is a d-dimensional martingale on I with $M_n(0) = 0$, and that for some continuous functions $\sigma^{ij}(t)$, the following holds for every fixed $t \in I$ as $n \to \infty$,*

$$\mathrm{E}\, M_n^i(t) M_n^j(t) \to \sigma^{ij}(t), \tag{2.6}$$

$$\mathrm{Var}([M_n^i, M_n^j]_t) \to 0. \tag{2.7}$$

Then $M_n \xrightarrow{d} M$ as $n \to \infty$, in the Skorokhod topology on $D(I)$, where M is a continuous d-dimensional Gaussian process with $\mathrm{E}\,M(t) = 0$ and covariance function

$$\mathrm{E}\, M^i(s) M^j(t) = \sigma^{ij}(s), \quad 0 \le s \le t < \infty. \tag{2.8}$$

Proof. The case $I = [0, \infty)$ is proved in [10] as a consequence of [8, Theorem VIII.3.12]. The case $I = [0, b]$, $b < \infty$, follows from this by considering the martingales $M_n(t \wedge b)$ on $[0, \infty)$. The remaining cases follow from these by continuous monotone transformations of the variable t. $\qquad\square$

Semimartingales. A stochastic process X defined on an interval $[a, b]$ or $[a, b)$ such that $X(t)$ is $\mathcal{F}_t$-measurable and

$$X(t) = M(t) + \int_a^t \xi(s)\,ds, \tag{2.9}$$

where M is a martingale and ξ is a stochastic process, is said to be a *semimartingale* with *drift* $\xi(t)$. (This is not the most general type of semimartingale, but it suffices for our purposes. In general, $M(t)$ only needs to be a local martingale, and the drift part may be any process of finite variation instead of $\int \xi(s)\,ds$.)

Itô's lemma shows that smooth functions of semimartingales are semimartingales and enables us to compute their drifts. We will only need a simple special case.

PROPOSITION 2.7. *If $X(t)$ and $Y(t)$ are semimartingales with drifts $\xi(t)$ and $\eta(t)$, and, further, processes of finite variation without common jumps, then $X(t)Y(t)$ is a semimartingale with drift $\xi(t)Y(t) + X(t)\eta(t)$.* $\qquad\square$

EXAMPLE 2.1. $\check{I}_e(t)$ is a martingale on $[0, 1)$, i.e. a semimartingale with drift 0, and the deterministic function $1 - t$ is trivially a semimartingale with drift -1. Hence $I_e'(t) = (1 - t)\check{I}_e(t)$ is a semimartingale with drift $-\check{I}_e(t)$, and $I_e(t) = I_e'(t) + t$ is a semimartingale on $[0, 1)$ with drift $1 - \check{I}_e(t) = (1 - I_e(t))/(1 - t)$.

Example 2.1 and Proposition 2.7 yield the following, which applies to $\psi(G_n(t))$ for any graph statistic ψ.

PROPOSITION 2.8. *If φ is a polynomial in the $\binom{n}{2}$ variables I_e, $e \in K_n$, which is linear in each I_e, then $\varphi(I_e(t))$ is a semimartingale on $[0, 1)$ with drift $(1 - t)^{-1} \sum_e \frac{\partial \varphi}{\partial I_e} \cdot (1 - I_e(t))$.* $\qquad\square$

We will use the following uniform estimates.

LEMMA 2.2. *Let $X(t)$ be a bounded semimartingale with drift $\xi(t)$. Then, for $s \leq u$,*

(i)
$$\mathrm{E} \sup_{s \leq t \leq u} |X(t)|^2 \leq \left(2\|X(u)\|_2 + 3 \int_s^u \|\xi(t)\|_2 dt \right)^2$$
$$\leq 13 \, \mathrm{E} \, |X(u)|^2 + 13 \left(\int_s^u \|\xi(t)\|_2 dt \right)^2$$

and

(ii) $\| \sup_{s \leq t \leq u} |X(t) - X(s)| \|_2$
$$\leq 2 \big(\mathrm{E}(X(u))^2 - \mathrm{E}(X(s))^2 \big)_+^{1/2} + 2\sqrt{2} \|X(u)\|_2^{1/2} \left(\int_s^u \|\xi(t)\|_2 \, dt \right)^{1/2}$$
$$+ 3 \int_s^u \|\xi(t)\|_2 dt.$$

Proof. Let $M(t) = X(t) - \int_s^t \xi(x)dx$. $M(t)$ is a martingale on $[s, u]$ and thus, because

$$\sup_{s \leq t \leq u} |X(t)| \leq \sup_{s \leq t \leq u} |M(t)| + \int_s^u |\xi(x)|dx, \qquad (2.10)$$

Doob's and Minkowski's inequalities yield

$$\| \sup |X(t)| \|_2 \leq \| \sup |M(t)| \|_2 + \| \int_s^u |\xi(t)|dt \|_2$$
$$\leq 2\|M(u)\|_2 + \int_s^u \|\xi(t)\|_2 dt$$
$$\leq 2\|X(u)\|_2 + 3 \int_s^u \|\xi(t)\|_2 dt, \qquad (2.11)$$

which, together with Cauchy-Schwarz' inequality, proves (i).

The same argument applied to $X(t) - X(s)$ yields

$$\| \sup |X(t) - X(s)| \|_2 \leq 2\|M(u) - M(s)\|_2 + \int_s^u \|\xi(t)\|_2 dt. \qquad (2.12)$$

Furthermore, if $\Xi = \int_s^u \xi(t)dt$,

$$\|M(u) - M(s)\|_2 = \big(\mathrm{E}(M(u) - M(s))^2 \big)^{1/2} = \big(\mathrm{E} \, M(u)^2 - \mathrm{E} \, M(s)^2 \big)^{1/2}$$
$$= \big(\mathrm{E}(X(u) - \Xi)^2 - \mathrm{E}(X(s))^2 \big)^{1/2}$$
$$= \big(\mathrm{E}(X(u))^2 - 2 \, \mathrm{E}(X(u)\Xi) + \mathrm{E}\,\Xi^2 - \mathrm{E}(X(s))^2 \big)^{1/2}$$
$$\leq \big(\mathrm{E}(X(u))^2 - \mathrm{E}(X(s))^2 + 2\|X(u)\|_2 \|\Xi\|_2 + \|\Xi\|_2^2 \big)^{1/2}$$
$$\leq \big(\mathrm{E}(X(u))^2 - \mathrm{E}(X(s))^2 \big)_+^{1/2} + \sqrt{2}\|X(u)\|_2^{1/2}\|\Xi\|_2^{1/2} + \|\Xi\|_2.$$
$$(2.13)$$

The estimate (ii) follows by (2.12), (2.13), and Minkowski's inequality. $\square$

Wick products. If $\xi_1, \ldots, \xi_m$ have a joint normal distribution with zero expectations, their Wick product is defined to be

$$:\!\xi_1 \cdots \xi_m\!: \; = \xi_1 \cdots \xi_m - \eta, \tag{2.14}$$

where η is a polynomial of degree at most $m - 1$ in $\xi_1, \ldots, \xi_m$, chosen such that $:\!\xi_1 \cdots \xi_m\!:$ is orthogonal to all such polynomials. Then

$$:\!\xi_1 \xi_2\!: \; = \xi_1 \xi_2 - \mathrm{E}(\xi_1 \xi_2), \tag{2.15}$$

$$:\!\xi_1 \xi_2 \xi_3\!: \; = \xi_1 \xi_2 \xi_3 - \mathrm{E}(\xi_2 \xi_3)\xi_1 - \mathrm{E}(\xi_1 \xi_3)\xi_2 - \mathrm{E}(\xi_1 \xi_2)\xi_3, \tag{2.16}$$

and, in general,

$$:\!\xi_1 \cdots \xi_m\!: \; = \sum (-1)^l \prod_{k=1}^{l} \mathrm{E}(\xi_{i_k} \xi_{j_k}) \prod_{r \in A} \xi_r \tag{2.17}$$

where we sum over all subsets A of $\{1, \ldots, m\}$ and partitions of $\{1, \ldots, m\} \setminus A$ into $l \geq 0$ pairs $\{i_k, j_k\}$.

In particular, $:\!\xi_1 \cdots \xi_m\!: \; = \xi_1 \cdots \xi_m$ if $\xi_1, \ldots, \xi_m$ are independent; on the other hand, if $\xi_1 = \cdots = \xi_m = \xi \sim N(0, 1)$, then $:\!\xi_1 \cdots \xi_m\!:$ is a Hermite polynomial in ξ.

We note the following well-known formula, where we sum over all permutations π of $1, \ldots, n$.

$$\mathrm{E}(:\!\xi_1 \cdots \xi_n\!: \; :\!\eta_1 \cdots \eta_m\!:) = \delta_{nm} \sum_{\pi} \prod_{i=1}^{n} \mathrm{E}(\xi_i \eta_{\pi(i)}). \tag{2.18}$$

3. The basic limit theorem

THEOREM 1. *Suppose that $(p_n)_1^{\infty}$ is a sequence of positive numbers such that $p_n \to p$ as $n \to \infty$, with $0 \leq p \leq 1$. Then there exist continuous stochastic processes $U(H; t)$, $t \geq 0$, indexed by unlabelled graphs, such that if H is any graph for which*

$$np_n^{m(H)} \to \infty, \tag{3.1}$$

then, as $n \to \infty$,

$$n^{-v(H)/2 - v_0(H)/2} p_n^{-e(H)/2} S_n(H; p_n t) \xrightarrow{d} U(H; t) \tag{3.2}$$

in the Skorokhod topology on $D([0, \infty))$. (In particular, (3.2) holds for every fixed $t \geq 0$.) The convergence in (3.2) holds jointly for any finite number of graphs H that satisfy (3.1). The limit processes are determined by the following properties.

(i) *If H is connected and $e(H) > 0$, then $U(H; t)$ is a Gaussian process with mean $\mathrm{E}\,U(H; t) = 0$ and covariance function*

$$\mathrm{E}\,U(H; s)U(H; t) = \mathrm{aut}(H)(s(1 - pt)_+)^{e(H)}, \qquad 0 \leq s \leq t < \infty. \tag{3.3}$$

(ii) If $H_1, \ldots, H_m$ are different connected unlabelled graphs, then $U(H_1; t)$, $\ldots, U(H_m; t)$ are independent processes.

(iii) If H has connected components $H_1, \ldots, H_k$, where $H_1, \ldots, H_m$ have at least one edge each while $H_{m+1}, \ldots, H_k$ consist of single vertices $(0 \leq m \leq k)$, then

$$U(H; t) = :U(H_1; t) \cdots U(H_m; t): . \tag{3.4}$$

Furthermore, (3.3) holds for any H with $v_0(H) = 0$, while $\mathrm{E}\, U(H_1; s)U(H_2; t) = 0$ if H_1 and H_2 are two different unlabelled graphs without isolated vertices.

REMARK 3.1. The assumption (3.1) is obviously equivalent to

$$np_n^{e(F)/v(F)} \to \infty, \qquad \emptyset \neq F \subseteq H \tag{3.5}$$

or, equivalently,

$$n^{v(F)}p_n^{e(F)} \to \infty, \qquad \emptyset \neq F \subseteq H. \tag{3.6}$$

It follows that (3.1) holds for a graph H if and only if it holds for all components of H.

REMARK 3.2. Formula (3.4) includes the special case $U(\emptyset; t) = 1$, and, more generally, $U(H; t) = 1$ for any discrete graph H. On the other hand, $\mathrm{E}\, U(H; t) = 0$ for any non-discrete H.

REMARK 3.3. The case $p_n = p = 1$ of Theorem 1 was given in [10]. Note that, in this case, (3.1) holds for any H. Furthermore, the processes $S_n(H; t)$ and $U(H; t)$ are 0 for $t \geq 1$ (except in the trivial case when H is discrete) so it suffices to study the processes on $[0, 1]$.

REMARK 3.4. The other main case is $p = 0$ (i.e. $p_n \to 0$); the case $0 < p \leq 1$ being a rather trivial extension of the case $p_n = 1$ (in fact, we may also allow $1 < p < \infty$ in Theorem 1). If $p = 0$, (3.3) simplifies to

$$\mathrm{E}\, U(H; s)U(H; t) = \mathrm{aut}(H)s^{e(H)}, \qquad 0 \leq s \leq t, \tag{3.7}$$

which means that, when H is connected and $e(H) > 0$, $U(H)$ is a Wiener process up to a (deterministic) change of time.

REMARK 3.5. The Wick product in (3.4) may be written using (multiple) stochastic integrals as

$U(H; t)$

$$= \sum_\pi \int_0^t \int_0^{t_1} \cdots \int_0^{t_{m-2}} U(H_{\pi(1)}; t_{m-1}) \, dU(H_{\pi(2)}; t_{m-1}) \cdots dU(H_{\pi(m)}; t_1), \tag{3.8}$$

where we sum over all permutations π of $1, \ldots, m$. This suggests that there may be a proof of the disconnected case of Theorem 1 using stochastic integrals, although we have not (yet) found it.

Proof. We will use Proposition 2.6. The following lemma contains the main combinational part of the argument.

LEMMA 3.1. *Suppose that G and H are connected graphs, and that (t_n) is a sequence with $\sup t_n < 1$ and $nt_n^{m(H)} \to \infty$. Then*

$$\operatorname{Var}([\check{S}_n(G), \check{S}_n(H)]_{t_n}) = o(n^{v(G)+v(H)} t_n^{e(G)+e(H)}). \tag{3.9}$$

Proof. The process $\check{S}_n(G;t)$ has a finite number of jumps at the times T_e, and it is obvious that $\check{S}_n(G;t)$ is of finite variation on every interval $[0,u]$, $u < 1$. Hence the quadratic covariation equals the sum of the products of the jumps;

$$[\check{S}_n(G), \check{S}_n(H)]_t = \sum_{T_e \leq t} \Delta\check{S}_n(G;T_e)\Delta\check{S}_n(H;T_e), \quad t < 1. \tag{3.10}$$

Furthermore, T_e are a.s. distinct and then

$$\Delta\check{S}_n(G;T_e) = (1 - T_e)^{-e(G)} \sum_{G_1} I(e \in G_1) \prod_{f \in G_1 \backslash \{e\}} I'_f(T_e), \tag{3.11}$$

where we sum over all copies G_1 of G in K_n, each copy counted $\operatorname{aut}(G)$ times. Thus, a.s.,

$$[\check{S}_n(G), \check{S}_n(H)]_t$$
$$= \sum_{e \in K_n} I(T_e \leq t)(1 - T_e)^{-e(G)-e(H)} \sum_{G_1, H_1} I(e \in G_1 \cap H_1)$$
$$\cdot \prod_{G_1 \backslash \{e\}} I'_f(T_e) \prod_{H_1 \backslash \{e\}} I'_f(T_e)$$
$$= \sum_{G_1, H_1} \sum_{e \in G_1 \cap H_1} I(T_e \leq t)(1 - T_e)^{-e(G)-e(H)} \prod_{G_1 \backslash \{e\}} I'_f(T_e) \prod_{H_1 \backslash \{e\}} I'_f(T_e)$$
$$= \sum_{G_1, H_1} \sum_{e \in G_1 \cap H_1} Y(G_1, H_1, e), \tag{3.12}$$

where $Y(G_1, H_1, e)$ denotes the summand in the preceding sum, and G_1 and H_1 range over the copies of G and H in K_n, counted with multiplicities. Consequently,

$$\operatorname{Var}([\check{S}_n(G), \check{S}_n(H)]_t) = \sum_{G_1, H_1, G_2, H_2} \sum_{\varepsilon_i \in G_i \cap H_i} \operatorname{Cov}(Y(G_1, H_1, \varepsilon_1), Y(G_2, H_2, \varepsilon_2)), \tag{3.13}$$

where G_i and H_i range over the copies of G and H, respectively, in K_n.

Consider one of the terms in this sum and let, for $j = 1,2,3,4$, v_j and e_j be the numbers of vertices and edges in $G_1 \cup H_1 \cup G_2 \cup H_2$ that belong to exactly j of G_1, H_1, G_2, H_2. We also write $v = v(G_1 \cup H_1 \cup G_2 \cup H_2) = \sum v_j$ and $e = e(G_1 \cup H_1 \cup G_2 \cup H_2) = \sum e_j$.

Suppose first that $e_1 \geq 1$, i.e. that there exists an edge f which belongs to exactly one of G_1, H_1, G_2, H_2. Since $\operatorname{E} I'_f(s) = 0$ for all s, and I'_f is independent of I'_ε and T_ε for all $\varepsilon \neq f$, it then follows, by first conditioning on T_{ε_1}

and T_{ε_2} and noting that $f \neq \varepsilon_1, \varepsilon_2$, that $E(Y(G_1, H_1, \varepsilon_1)Y(G_2, H_2, \varepsilon_2)) = 0 = E Y(G_1, H_1, \varepsilon_1) E Y(G_2, H_2, \varepsilon_2)$, and consequently the covariance vanishes.

It is obvious that if $e_1 = 0$ then $v_1 = 0$ too. Furthermore, if $e_1 = v_1 = v_3 = v_4 = 0$, then each vertex in one of G_1, H_1, G_2, H_2 belongs to exactly 2 of them. Since the graphs are connected and $e_1 = 0$, this is possible only if the graphs coincide in two disjoint pairs, and since $G_1 \cap H_1$ contains ε_1 and thus is non-empty, $G_1 = H_1$ and $G_2 = H_2$ with $G_1 \cap G_2 = \emptyset$. This implies, however, that $Y(G_1, H_1, \varepsilon_1)$ and $Y(G_2, H_2, \varepsilon_2)$ are independent, and again the covariance vanishes. Consequently we only have to consider terms with $e_1 = v_1 = 0$ and $v_3 + v_4 \geq 1$. Let $t_0 = \sup t_n < 1$. Since, for $j_1 + j_2 \geq 1$, $E I'_f(s_1)^{j_1} I'_f(s_2)^{j_2} = O(s_1 \vee s_2)$, we have for $t \leq t_0$, for some $C(t_0) < \infty$,

$$| E(Y(G_1, H_1, \varepsilon_1)Y(G_2, H_2, \varepsilon_2)) \mid T_{\varepsilon_1}, T_{\varepsilon_2}|$$
$$\leq C(t_0)I(T_{\varepsilon_1}, T_{\varepsilon_2} \leq t)(T_{\varepsilon_1} \vee T_{\varepsilon_2})^{e(G_1 \cup H_1 \cup G_2 \cup H_2 \setminus \{\varepsilon_1, \varepsilon_2\})} \quad (3.14)$$

and thus (considering the cases $\varepsilon_1 = \varepsilon_2$ and $\varepsilon_1 \neq \varepsilon_2$ separately)

$$E(Y(G_1, H_1, \varepsilon_1)Y(G_2, H_2, \varepsilon_2)) = O(t^{e(G_1 \cup H_1 \cup G_2 \cup H_2)}), \qquad t \leq t_0. \quad (3.15)$$

This together with similar estimates for $E Y(G_i, H_i, \varepsilon_i)$ yields

$$\text{Cov}(Y(G_1, H_1, \varepsilon_1), Y(G_2, H_2, \varepsilon_2)) = O(t^e), \qquad t \leq t_0. \quad (3.16)$$

Since the same configuration of G_1, H_1, G_2, H_2 occurs $O(n^v)$ times in the sum (3.13), and the total number of configurations is finite, it only remains to prove that, assuming $e_1 = v_1 = 0$ and $v_3 + v_4 \geq 1$,

$$n^v t_n^e = o(n^{v(G)+v(H)} t_n^{e(G)+e(H)}). \quad (3.17)$$

In order to see this, we observe that

$$\sum_1^4 j v_j = v(G_1) + v(H_1) + v(G_2) + v(H_2) = 2v(G) + 2v(H) \quad (3.18)$$

and, similarly,

$$\sum_1^4 j e_j = 2e(G) + 2e(H). \quad (3.19)$$

Hence

$$n^{2v} t_n^{2e} n^{-2v(G)-2v(H)} t_n^{-2e(G)-2e(H)} = n^{2\sum v_j - \sum j v_j} t_n^{2\sum e_j - \sum j e_j}$$
$$= n^{-v_3 - 2v_4} t_n^{-e_3 - 2e_4}. \quad (3.20)$$

Now define $H' = G_1 \cap H_1 \cap G_2$ and $H'' = ((G_1 \cap H_1) \cup (G_1 \cap G_2) \cup (H_1 \cap G_2)) \cap H_2$. It is easy to check that

$$v(H') + v(H'') = v_3 + 2v_4, \quad (3.21)$$
$$e(H') + e(H'') = e_3 + 2e_4, \quad (3.22)$$

and thus

$$n^{v_3+2v_4} t_n^{e_3+2e_4} = n^{v(H')} t_n^{e(H')} n^{v(H'')} t_n^{e(H'')}. \qquad (3.23)$$

Since $H' \subseteq H_1$, the assumption $n t_n^{m(H)} \to \infty$ implies that $n^{v(H')} t_n^{e(H')} \to \infty$ unless $H' = \emptyset$ (cf. Remark 3.1), and similarly $n^{v(H'')} t_n^{e(H'')} \to \infty$ unless $H'' = \emptyset$. At least one of H' and H'' has to be nonempty, however, because otherwise $v_3 + 2v_4 = 0$ by (3.21), contrary to our assumption. Consequently, (3.23) yields $n^{v_3+2v_4} t_n^{e_3+2e_4} \to \infty$, which by (3.20) proves (3.17), and the proof of the lemma is completed. $\qquad\Box$

It will be convenient to have a name for the left hand side of (3.2), so we define

$$U_n(H;t) = n^{-v(H)/2 - v_0(H)/2} p_n^{-e(H)/2} S_n(H; p_n t). \qquad (3.24)$$

We also define, for $p_n t < 1$,

$$\begin{aligned}\check{U}_n(H;t) &= (1 - p_n t)^{-e(H)} U_n(H;t) \\ &= n^{-v(H)/2 - v_0(H)/2} p_n^{-e(H)/2} \check{S}_n(H; p_n t);\end{aligned} \qquad (3.25)$$

by Lemma 2.1, $\check{U}_n(H)$ is a martingale on $[0, 1/p_n)$.

Now consider a finite family Γ of connected unlabelled graphs that all have at least two vertices and satisfy (3.1). Fix $t_0 > 0$ with $pt_0 < 1$ and consider only n that are so large that $p_n t_0 < 1$. Then the processes $\check{U}_n(H)$, $H \in \Gamma$, together form a vector-valued martingale on $[0, t_0]$. If $G, H \in \Gamma$ and $t \in (0, t_0]$ is fixed, Lemma 3.1 with $t_n = p_n t$ yields

$$\mathrm{Var}([\check{U}_n(G), \check{U}_n(H)]_t) \to 0. \qquad (3.26)$$

Furthermore, by Proposition 1.1,

$$\begin{aligned}\mathrm{Var}(\check{U}_n(H;t)) &= n^{-v(H)} p_n^{-e(H)} (1 - p_n t)^{-2e(H)} \mathrm{aut}(H)(n)_{v(H)} (p_n t(1 - p_n t))^{e(H)} \\ &\to \mathrm{aut}(H)(1 - pt)^{-e(H)} t^{e(H)},\end{aligned} \qquad (3.27)$$

while, if $G \neq H$,

$$\mathrm{Cov}(\check{U}_n(G;t), \check{U}_n(H;t)) = 0. \qquad (3.28)$$

The martingale convergence theorem Proposition 2.6 thus applies and yields $\check{U}_n(H) \xrightarrow{d} \check{U}(H)$ in $D([0, t_0])$, jointly for $H \in \Gamma$, where $\check{U}(H)$ is a continuous Gaussian process such that $\mathrm{E}\,\check{U}(H;t) = 0$ and

$$\mathrm{E}\,\check{U}(H;s)\check{U}(H;t) = \mathrm{aut}(H)(1 - ps)^{-e(H)} s^{e(H)}, \qquad 0 \le s \le t \le t_0, \qquad (3.29)$$

and, further, $\check{U}(H)$ and $\check{U}(G)$ are uncorrelated, and thus independent, when $H \neq G$. Since $(1 - p_n t)^{e(H)} \to (1 - pt)^{e(H)}$ uniformly on $[0, t_0]$, this implies by Proposition 2.2

$$U_n(H;t) = (1 - p_n t)^{e(H)} \check{U}_n(H;t) \xrightarrow{d} (1 - pt)^{e(H)} \check{U}(H;t) \qquad (3.30)$$

in $D([0, t_0])$, jointly for $H \in \Gamma$.

We now define $U(H; t) = (1 - pt)^{e(H)} \check{U}(H; t)$. Then (3.30) shows that (3.2) holds in $D([0, t_0])$ for $H \in \Gamma$, and (3.3) follows from (3.29). The other assertions in Theorem 1(i),(ii) follow immediately, and we have proved the theorem for connected graphs containing at least two vertices, on $[0, t_0]$ for any $t_0 < 1/p$.

If $p = 0$ we can choose any finite t_0. It follows that there exist continuous Gaussian processes $U(H; t)$ on $[0, \infty)$ (for H connected with $v(H) \geq 2$), such that (3.3) is satisfied and the convergence (3.2) holds in every $D([0, t_0])$ and thus, by Proposition 2.3(ii), in $D([0, \infty))$, which proves the theorem for such H. If $p > 0$, the same argument yields the existence of $U(H; t)$ and (3.2) on $[0, 1/p]$; we leave the extension to the whole half-line $[0, \infty)$ until the end of the proof.

In order to prove the assertions for disconnected H, we will use two further lemmas.

LEMMA 3.2. *For any H, there exist $C = C(H) < \infty$ such that for all n and $t_0 > 0$,*

$$\mathrm{E}(\sup_{0 \leq t \leq t_0} |S_n(H; t)|)^2 \leq C n^{v(H) + v_0(H)} t_0^{e(H)}. \tag{3.31}$$

and thus

$$\mathrm{E}(\sup_{0 \leq t \leq t_0} |U_n(H; t)|)^2 \leq C t_0^{e(H)}. \tag{3.32}$$

Proof. We may, by Proposition 1.1(i), assume that $v_0(H) = 0$. If $t_0 < 1$, then Doob's inequality and Proposition 1.1(iii) yield

$$\mathrm{E}(\sup_{t \leq t_0} |S_n(H; t)|)^2 \leq \mathrm{E}(\sup_{t \leq t_0} |\check{S}_n(H; t)|)^2 \leq 4\,\mathrm{E}\,|\check{S}_n(H; t_0)|^2$$

$$= 4\,\mathrm{aut}(H)(1 - t_0)^{-e(H)}(n)_{v(H)} t_0^{e(H)}. \tag{3.33}$$

This proves (3.31) for $t_0 \leq 1/2$. For larger t_0 we use a time reversal. We may replace T_e by $1 - T_e$ for every e; this sends $I_e(t)$ into $1 - I_e(1 - t)$, $I'_e(t)$ into $-I'_e(1 - t)$ and $S_n(H; t)$ into $(-1)^{e(H)} S_n(H; 1 - t)$ (adjusted to be right-continuous at the jumps). Hence $\sup_{1/2 \leq t \leq 1} |S_n(H; t)|$ and $\sup_{0 \leq t \leq 1/2} |S_n(H; t)|$ have the same distribution, and the case just proved shows that

$$\mathrm{E}(\sup_{0 \leq t < \infty} |S_n(H; t)|)^2 \leq \mathrm{E}(\sup_{t \leq 1/2} |S_n(H; t)|)^2 + \mathrm{E}(\sup_{1/2 \leq t \leq 1} |S_n(H; t)|)^2$$

$$\leq C(H) n^{v(H)}.$$

Consequently (3.31) holds for $t_0 > 1/2$ as well; (3.32) now follows by (3.24). □

LEMMA 3.3. *Suppose that H is a graph with components $H_1, \ldots, H_m$ and that H' is a connected graph, and furthermore that $v_0(H) = v_0(H') = 0$. If (3.1) holds and $t_0 < 1/p$, then, as $n \to \infty$,*

$$\mathrm{E}\bigg(\sup_{0 \leq t \leq t_0} |U_n(H + H'; t) - U_n(H; t) U_n(H'; t)$$

$$+ \sum_{H_i \cong H'} \mathrm{aut}(H')(t(1 - p_n t))^{e(H')} U_n(H - H_i; t)|\bigg)^2 \to 0. \tag{3.34}$$

Proof. By the definition (1.3) of S_n,

$$S_n(H;t)S_n(H';t) = \sum_{\varphi,\varphi'} \prod_{e\in\varphi(H)} I'_e(t) \prod_{e\in\varphi'(H')} I'_e(t), \qquad (3.35)$$

where φ and φ' range over the injective mappings from $V(H)$ and $V(H')$ into $\{1,\ldots,n\}$. By grouping the terms in (3.35) according to the way $\varphi(H)$ and $\varphi'(H')$ intersect, and expanding all squares that appear, using $(I'_e(t) - (1 - t))(I'_e(t) + t) = 0$ and thus $I'_e(t)^2 = (1 - 2t)I'_e(t) + t(1 - t)$, we obtain

$$S_n(H;t)S_n(H';t) = \sum (1 - 2t)^{e(F\cap F_2)}(t(1 - t))^{e(F_1)-e(F)}S_n(F;t), \qquad (3.36)$$

where we sum over all F_1, F_2, F such that for some $H'_1 \cong H'$, $F_1 = H \cup H'_1$, $F_2 = H \cap H'_1$ and $V(F) = V(F_1)$, $E(F_1) \setminus E(F_2) \subseteq E(F) \subseteq E(F_1)$, and we consider only one H'_1 (and thus one F_1, F_2) for every identification of $d \geq 0$ vertices in $V(H)$ with d vertices in $V(H')$. The single term on the right hand side of (3.36) with H and H'_1 disjoint, and thus $F_2 = \emptyset$, $F = F_1 = H + H'$, equals $S_n(H + H';t)$. (This is given by all terms in (3.35) with $\varphi(H)$ and $\varphi'(H')$ disjoint.) There is only a finite number of other terms in (3.36), independently of n (in fact, there are exactly $\sum_1^{v(H)} \binom{v(H)}{d}\binom{v(H')}{d}d!$ of them) and we estimate them separately. Let $X_n(t)$ be one of them. Then, by Lemma 3.2, with C_1 depending on F and t_0,

$$\mathrm{E}(\sup_{t\leq t_0} |X_n(p_n t)|)^2 \leq C(p_n t_0)^{2(e(F_1)-e(F))}n^{v(F)+v_0(F)}(p_n t_0)^{e(F)}$$

$$= C_1 n^{v(H)+v(H')-v(F_2)+v_0(F)}p_n^{e(H)+e(H')-e(F_2)+e(F_1)-e(F)}$$

$$= C_1 n^{v(H)+v(H')-v(F^*)}p_n^{e(H)+e(H')-e(F^*)} \qquad (3.37)$$

where F^* is the graph $F \cap F_2$ with all isolated vertices in F deleted (these vertices have to belong to F_2, since neither H nor H' contains isolated vertices). If $F^* \neq \emptyset$, then $n^{v(F^*)}p_n^{e(F^*)} \to \infty$ by (3.6), because $F^* \subseteq F_2 \subseteq H$, and thus (3.37) yields

$$\mathrm{E}(\sup_{t\leq t_0} |X_n(p_n t)|)^2 = o(n^{v(H)+v(H')}p_n^{e(H)+e(H')}).$$

If $F^* = \emptyset$, every vertex in $V(F_2) = V(H) \cap V(H'_1)$ is isolated in F, and (since H' is assumed to be connected) this happens only if either $F_2 = \emptyset$, the case already treated, or $F_2 = H'_1$ equals a component H_i of H, and $E(F) = E(F_1) \setminus E(F_2)$. In the latter case, the term $X_n(t)$ equals

$$(t(1 - t))^{e(H')}S_n(F;t) = (t(1 - t))^{e(H')}(n - (v(H) - v(H')))_{v(H')}S_n(H - H_i;t),$$

so by Lemma 3.2 again

$$\mathrm{E}\big(\sup_{t\leq t_0} |X_n(p_n t) - n^{v(H')}p_n^{e(H')}(t(1 - p_n t))^{e(H')}S_n(H - H_i;p_n t)|\big)^2$$

$$= O\big(n^{2(v(H')-1)}p_n^{2e(H')}t_0^{2e(H')}n^{v(H)-v(H')}(p_n t_0)^{e(H)-e(H')}\big)$$

$$= o(n^{v(H)+v(H')}p_n^{e(H)+e(H')}). \qquad (3.38)$$

Since there are $\mathrm{aut}(H')$ such terms for every $H_i \cong H'$, the conclusion follows from these estimates and (3.36), upon normalization with the factor $n^{-v(H)/2-v(H')/2}p_n^{-e(H)/2-e(H')/2}$. $\qquad\square$

We can now complete the proof of the theorem. Let H be a disconnected graph with components $H_1, \ldots, H_m$ and define $U(H;t)$ by (3.4). Then $U(H;t)$ is a polynomial in $U(H_i;t)$ and is thus a continuous stochastic process on $[0, 1/p)$. The final assertion in the theorem follows from (3.3) and (2.18). We prove that $U_n(H;t) \to U(H;t)$ in $D([0,1/p))$ by induction on m, the number of components of H; by Proposition 1.1(i) we may assume that $v_0(H) = 0$. The case $m = 0$ ($H = \emptyset$) is trivial and $m = 1$ is proved above. If $m \geq 2$, Lemma 3.3 shows that $U_n(H;t) - U_n(\sum_1^{m-1} H_i;t)U_n(H_m;t) + \sum_{H_i \cong H_m} \mathrm{aut}(H_m)t(1 - p_nt)^{e(H_m)}U_n(\sum_{j\neq i,m} H_j;t) \to 0$ in probability in $D([0,t_0])$ for every $t_0 < 1/p$, and thus in $D([0,1/p))$. Consequently, by induction, Proposition 2.2 and (2.17),

$$U_n(H;t) = U_n(\sum_1^m H_i;t)$$

$$\xrightarrow{d} U(\sum_1^{m-1} H_i;t)U(H_m;t) - \sum_{i=1}^{m-1} I(H_i \cong H_m)\,\mathrm{aut}(H_m)(t(1-pt))^{e(H_m)}$$
$$\cdot U(\sum_{j\neq i,m} H_j;t)$$

$$= :U(H_1;t)\cdots U(H_{m-1};t):U(H_m;t)$$

$$- \sum_{i=1}^{m-1} \mathrm{E}(U(H_i;t)U(H_m;t)):U(H_1;t)\cdots U(H_{i-1};t)U(H_{i+1};t)\cdots U(H_{m-1};t):$$

$$= :U(H_1;t)\cdots U(H_m;t):, \qquad\qquad (3.39)$$

with convergence in $D([0,1/p))$, jointly in any finite number of graphs. This proves the theorem on $[0,1/p)$; in particular, the proof is completed for the case $p = 0$.

If $p > 0$, assume $e(H) > 0$ (otherwise the result is trivial), and define $U(H;t) = 0$, $t \geq 1/p$. This satisfies the claims in (i)–(iii) in the statement of the theorem. We have shown that $U(H)$ is continuous on $[0,1/p)$, and since $U(H;t)$ and $U(H;1/p-t)$, $0 \leq t \leq 1/p$, have the same distribution, $U(H)$ is a.s. continuous also at $1/p$ and thus everywhere.

The time symmetry yields also, together with Lemma 3.2,

$$\mathrm{E}\sup_{t\geq u}|U_n(H;t)|^2 = \mathrm{E}\sup_{0\leq t\leq (1/p_n-u)_+}|U_n(H;t)|^2 \leq C(\tfrac{1}{p_n}-u)_+^{e(H)}. \qquad (3.40)$$

Consequently,

$$\limsup_{n\to\infty} \mathrm{E}\Big(\sup_{t\geq u}|U_n(H;t)|^2\Big) \leq C(\tfrac{1}{p}-u)_+^{e(H)} \to 0 \qquad \text{as } u \to 1/p, \qquad (3.41)$$

which by Proposition 2.4 shows that $U_n(H;t) \xrightarrow{d} U(H;t)$ in $D([0,1/p])$.

Furthermore, (3.40) implies that $\mathrm{E}(\sup_{t \geq 1/p} |U_n(H;t)|)^2 \to 0$ and thus $U_n(H)$ $\xrightarrow{p} U(H) = 0$ in $D([1/p, \infty))$. Hence, using [4, Theorem 4.4], $U_n(H) \xrightarrow{d} U(H)$ in $D([0, 1/p])$ and in $D([1/p, \infty))$, jointly, and thus, by Proposition 2.3(iii), also in $D([0, \infty))$.

Joint convergence for several H follows in the same way (use vector valued versions of Propositions 2.4 and 2.3). $\qquad\square$

REMARK 3.6. The proof shows that if $p > 0$, then (3.2) holds also in $D([0, \infty])$ with $U_n(H; \infty) = U(H; \infty) = 0$.

4. The orthogonal decomposition

We begin by a formal statement of the orthogonal decomposition described in the introduction. Recall that $\mathcal{U}^0$ is the set of all unlabelled graphs without isolated vertices.

PROPOSITION 4.1. If ψ is any graph statistic, then $\psi(G_n(t))$ has an orthogonal decomposition

$$\psi(G_n(t)) = \sum_{H \in \mathcal{U}^0} \hat{\psi}_n(H;t) S_n(H;t) \tag{4.1}$$

where each $\hat{\psi}_n(H;t)$ is a continuous function of t. (In fact, $\hat{\psi}_n(H;t)$ is a polynomial in t for $0 \leq t \leq 1$.) Furthermore,

$$\mathrm{Var}(\psi(G_n(t))) = \sum_{H \in \mathcal{U}^0} \hat{\psi}_n(H;t)^2 \, \mathrm{Var}(S_n(H;t))$$

$$= \sum_{H \neq \emptyset} \hat{\psi}_n(H;t)^2 \, \mathrm{aut}(H)(n)_{v(H)} \big(t(1-t)\big)^{e(H)}, \qquad 0 \leq t \leq 1. \tag{4.2}$$

Proof.
$$\psi(G_n(t)) = \sum_G \psi(G) \prod_{e \in G} I_e(t) \prod_{e \notin G} (1 - I_e(t)), \tag{4.3}$$

where we sum over all graphs G with vertex set $\{1, \ldots, n\}$. We may assume $0 \leq t \leq 1$, and substitute $I_e(t) = I'_e(t) + t$. After expansion, $\psi(G_n(t))$ will be expressed as a linear combination of terms $\prod_{e \in H} I'_e(t)$, $H \subseteq K_n$, and (4.1) is obtained by collecting terms with isomorphic H together. (4.2) follows immediately from Proposition 1.1. $\qquad\square$

The decomposition in Proposition 4.1 is unique. In fact, the orthogonality (Proposition 1.1 (iv)) yields, unless $S_n(H;t) = 0$ a.s.,

$$\hat{\psi}_n(H;t) = \mathrm{E}\big(\psi(G_n(t)) S_n(H;t)\big) / \mathrm{E}\, S_n(H;t)^2, \tag{4.4}$$

and thus the term $\hat{\psi}_n(H;t) S_n(H;t)$ is uniquely determined. We set, for definiteness, $\psi_n(H;t) = \psi_n(H;1)$ for $t > 1$ and $\psi_n(H;t) = 0$ when $v(H) > n$.

REMARK 4.1. We will only use H without isolated vertices in the decomposition. We leave $\hat{\psi}_n(H;t)$ undefined when H has an isolated vertex (or define it to be 0).

We define a (partly) normalized version by

$$\hat{\psi}_n^*(H;t) = n^{v(H)/2}t^{e(H)/2}\hat{\psi}(H;t). \tag{4.5}$$

Thus

$$\mathrm{E}(\hat{\psi}_n(H;t)S_n(H;t))^2 = |\hat{\psi}_n^*(H;t)|^2(1+O(\tfrac{1}{n}))(1-t)^{e(H)}\,\mathrm{aut}(H)$$
$$\asymp |\hat{\psi}_n^*(H;t)|^2, \tag{4.6}$$

provided H is fixed and $t \leq t_0 < 1$. Consequently, the most important terms in (4.1), for a given t, are those with largest $|\hat{\psi}_n^*(H;t)|$.

We also let $\psi_n^{\mathcal{H}}(t)$, where $\mathcal{H} \subseteq \mathcal{U}^0$, denote the projection of $\psi(G_n(t))$ onto the span of $\{S_n(H;t)\}_{H\in\mathcal{H}}$, i.e.,

$$\psi_n^{\mathcal{H}}(t) = \sum_{H\in\mathcal{H}} \hat{\psi}_n(H;t)S_n(H;t). \tag{4.7}$$

(In general, $\psi_n^{\mathcal{H}}$ is not a function of $G_n(t)$; it depends also explicitly on t.)

REMARK 4.2. If ψ_n denotes the restriction of ψ to graphs of order n, then $\hat{\psi}_n(H;t)$ depends only on ψ_n, for any given n. Conversely, the coefficients $\{\hat{\psi}_n(H;t)\}_{H\in\mathcal{U}^0}$ for a fixed $t \in (0,1)$ determine by (4.1) $\psi(G_n(t))$ and hence ψ_n, because $G_n(t)$ equals any given graph on the vertices $\{1,\dots,n\}$ with positive probability.

In particular, the values of the coefficients $\hat{\psi}_n(H;t)$ at a given $t \in (0,1)$ determine their values at any other t. (This also follows from repeated use of Proposition 4.3 below.)

On the other hand, if $t \in (0,1)$ is fixed, the coefficients $\hat{\psi}_n(H;t)$ may be arbitrarily specified. This follows either by construction ($S_n(H;t) = \psi(G_n(t))$ for a graph statistic $\psi = \psi_{H,t}$) or by noting that the linear mapping $\psi_n \to \{\hat{\psi}_n(H;t)\}_{v(H)\leq n}$ is an injection between two vector spaces of the same finite dimension $\#\{G \in \mathcal{U} : v(G) = n\} = \#\{H \in \mathcal{U}^0 : v(H) \leq n\}$, and thus is an isomorphism.

The term in (4.1) with $H = \emptyset$ is special since it is non-random; in fact it is just the expectation of the sum. The next coefficient, with $H = K_2$ is related to the derivative of the expectation, while higher derivatives can be obtained from higher coefficients.

PROPOSITION 4.2. If $0 < t < 1$, then

$$\mathrm{E}\,\psi(G_n(t)) = \hat{\psi}_n(\emptyset;t) \tag{4.8}$$

$$\frac{d}{dt}\,\mathrm{E}\,\psi(G_n(t)) = n(n-1)\hat{\psi}_n(K_2;t), \tag{4.9}$$

$$\frac{d^2}{dt^2}\,\mathrm{E}\,\psi(G_n(t)) = 2(n)_3\hat{\psi}_n(P_2;t) + 2(n)_4\hat{\psi}_n(K_2 + K_2;t). \tag{4.10}$$

Proof. Formula (4.8) follows from (4.1) and Proposition 1.1(ii). In order to prove (4.9) and (4.10) let us again use the expansion (4.3). Then $\mathrm{E}\,\psi(G_n(t))$ is obtained by replacing $I_e(t)$ by $\mathrm{E}\,I_e(t) = t$, so $\mathrm{E}\,\psi(G_n(t+\varepsilon))$ is obtained by replacing $I_e(t)$ by $t+\varepsilon$, i.e. by replacing $I'_e(t)$ by ε (assuming $0 < t+\varepsilon < 1$). Hence, comparing (4.1) and (1.3), we obtain

$$\mathrm{E}\,\psi(G_n(t+\varepsilon)) = \sum_{\mathcal{H}} \hat{\psi}_n(H;t)(n)_{v(H)}\varepsilon^{e(H)}. \tag{4.11}$$

Since $e(H) = 1$ only for $H = K_2$, (4.9) follows, while the two terms with $e(H) = 2$ yield (4.10).   □

A similar argument yields the following generalization of (4.9).

PROPOSITION 4.3.

$$\frac{d}{dt}\hat{\psi}_n(H;t) = \sum \binom{n - v(H)}{v(H_1) - v(H)}\frac{\mathrm{aut}(H_1)}{\mathrm{aut}(H)}\hat{\psi}_n(H_1;t), \qquad 0 < t < 1, \tag{4.12}$$

where we sum over the graphs H_1 (not necessarily different) obtained by adding an edge to H, either between two vertices of H ($\binom{v(H)}{2} - e(H)$ terms), between a vertex of H and a new vertex ($v(H)$ terms), or between 2 new vertices (1 term, viz. $H + K_2$).

Proof. Let $\mathrm{sub}(F,G)$ denote the number of subgraphs of G that are isomorphic to F. By linearity and (4.3), it suffices to consider the case $\psi(G) = \mathrm{sub}(F,G)$ for a fixed graph F. Then

$$\psi(G_n(t)) = \sum_H \frac{\mathrm{sub}(F,K_n)}{(n)_{v(H)}}\,\mathrm{sub}(H,F)t^{e(F)-e(H)}S_n(H;t), \tag{4.13}$$

and thus

$$\hat{\psi}_n(H;t) = \frac{\mathrm{sub}(F,K_n)}{(n)_{v(H)}}\,\mathrm{sub}(H,F)t^{e(F)-e(H)} \tag{4.14}$$

and

$$\frac{d}{dt}\hat{\psi}_n(H;t) = \frac{\mathrm{sub}(F,K_n)}{(n)_{v(H)}}\,\mathrm{sub}(H,F)(e(F) - e(H))t^{e(F)-e(H)-1}. \tag{4.15}$$

The assertion (4.12) follows because

$$\mathrm{aut}(H)\,\mathrm{sub}(H,F)(e(F) - e(H)) = \sum \frac{\mathrm{aut}(H_1)}{(v(H_1) - v(H))!}\,\mathrm{sub}(H_1,F), \tag{4.16}$$

the left hand side being the number of labelled subgraphs of F that are isomorphic to a given labelling of H together with an edge in F not in the subgraph.   □

The term in (4.1) with $H = K_2$ can be interpreted as the linear variation in ψ caused by the random variation of the number of edges in $G_n(t)$ (by orthogonality and $e(G_n(t)) = \frac{1}{2}S_n(K_2;t) + \binom{n}{2}t$); it also equals the *first projection*, which usually is defined as $\sum_{e\in K_n}\mathrm{E}\big(\psi(G_n(t)) - \mathrm{E}\,\psi(G_n(t)) \mid I_e(t)\big)$.

PROPOSITION 4.4. *For any ψ, the first projection equals $\psi_n^{\{K_2\}}(t)$.*

Proof. If $e(H) \geq 2$, or if $e(H) = 1$ and $e \notin H$, then $\mathrm{E}\big(\prod_{f \in H} I_f'(t) \mid I_e(t)\big) = 0$ by independence.

Consequently, $\mathrm{E}\big(S_n(H;t) \mid I_e(t)\big) = 0$ if $e(H) \geq 2$, and $\mathrm{E}\big(S_n(K_2;t) \mid I_e(t)\big) = 2I_e'(t)$, and it follows from (4.1) and (1.3) that the first projection equals

$$\sum_{e \in K_n} \hat{\psi}(K_2;t)2I_e'(t) = \hat{\psi}(K_2;t)S_n(K_2;t). \qquad \square$$

For each n, it is possible to use a finite decomposition in (4.1), because $S_n(H;t) = 0$ when $v(H) > n$ (or by the proof above), but for many natural statistics ψ, the required number of terms increase with n. We say that ψ has a *finite decomposition* if there is a finite set $\mathcal{H}$ which can be used for every n. This corresponds to the notion of an induced property, see [17]. For example, the number of k-subsets having a given induced property (k is fixed) is a statistic having a finite decomposition (only H with $v(H) \leq k$ are needed in (4.1)). We also define the *support* of ψ as $\{H \in \mathcal{U}^0 : \hat{\psi}_n(H;t) \neq 0 \quad$ for some $n,t\}$. Thus ψ has a finite decomposition if and only if its support is finite. We give examples of finite decompositions in Section 10 and non-finite decompositions in Section 11. A similar, but more complicated, example is treated in [2].

We note that having a finite decomposition severely restricts the non-zero coefficients $\hat{\psi}_n(H;t)$ too.

PROPOSITION 4.5. *If ψ has a finite decomposition, then each $\hat{\psi}_n(H;t)$ is a polynomial in t ($0 \leq t \leq 1$) of degree uniformly bounded in n and H.*

Proof. Let $\mathcal{H}$ be the support of ψ and $d = \max\{e(H) : H \in \mathcal{H}\} < \infty$. Then, by Proposition 4.3 and backward induction,

$$\deg(\hat{\psi}_n(H;t)) \leq d - e(H) \leq d. \qquad (4.17)$$

$$\square$$

If ψ has a finite decomposition, (4.1) and Theorem 1 immediately give limit theorems for $\psi(G_{n,p})$ as well as functional limit theorems, see Sections 5 and 8, and it is also easy to obtain limit theorems for $\psi(G_{n,m})$, see Section 6.

Furthermore, it is often possible to treat statistics that do not have finite decompositions by making suitable approximations. We introduce some relevant definitions.

We say that the decomposition (4.1) is *dominated* by a subfamily $\mathcal{H} \subseteq \mathcal{U}^0$, for a given sequence p_n, if

$$\mathrm{Var}\big(\psi_n^{\mathcal{U}^0 \setminus \mathcal{H}}(p_n)\big) = o\big(\mathrm{Var}(\psi(G_{n,p_n}))\big) \qquad (4.18)$$

or, equivalently, because of the orthogonality,

$$\mathrm{Var}\big(\psi_n^{\mathcal{H}}(p_n)\big) \sim \mathrm{Var}\big(\psi(G_{n,p_n})\big). \qquad (4.19)$$

If, furthermore, $\mathcal{H}$ is finite, we say that the decomposition is *finitely dominated*. It is then, again, easy to obtain limit theorems for $\psi(G_{n,p_n})$ from Theorem 1, and we will later see how results for $G_{n,m}$ can be obtained.

EXAMPLE 4.1. The method of the first projection, which has been used by several authors, works as follows. Let $X_n = \psi(G_{n,p_n})$ and let X'_n be the first projection defined above (cf. Proposition 4.4). Since X'_n is a linear function of a $\mathrm{Bi}\left(\binom{n}{2}, p_n\right)$ variable, X'_n is asymptotically normally distributed, provided only $n^2 p_n, n^2(1 - p_n) \to \infty$. If $\mathrm{Var}(X_n - X'_n) = o(\mathrm{Var}\, X_n)$, it follows that X_n too has an asymptotically normal distribution. Proposition 4.4 shows that this is the case precisely when the decomposition of ψ is dominated by $\{K_2\}$, so this is the simplest case of a finitely dominated decomposition and the normal limit is a special case of Theorem 3 in the next section.

The condition of being finitely dominated can be somewhat generalized as follows. We say that the decomposition is *almost finitely dominated* by $\mathcal{H} \subseteq \mathcal{U}^0$, for a given sequence p_n, if for every $\varepsilon > 0$, there exists a finite $\mathcal{H}_\varepsilon \subseteq \mathcal{H}$ such that, for all large n,

$$\mathrm{Var}\left(\psi_n^{\mathcal{U}^0 \setminus \mathcal{H}_\varepsilon}(p_n)\right) \leq \varepsilon\, \mathrm{Var}\left(\psi(G_{n,p_n})\right), \tag{4.20}$$

or, equivalently,

$$\mathrm{Var}\left(\psi_n^{\mathcal{H}_\varepsilon}(p_n)\right) \geq (1 - \varepsilon)\, \mathrm{Var}\left(\psi(G_{n,p_n})\right). \tag{4.21}$$

It follows immediately that (4.18) holds, so 'almost finitely dominated' implies 'dominated' (but not conversely).

We say that the statistic ψ or the variable $\psi(G_n(p_n))$ is (almost) finitely dominated when the corresponding decomposition is. Note that if the decomposition is not finite, it depends in general on the sequence p_n whether the decomposition is finitely dominated, almost finitely dominated, or neither, see e.g. the example in Section 11.

PROPOSITION 4.6. *If ψ is almost finitely dominated by some family, then ψ is almost finitely dominated by every family that dominates it, and in particular by*

$$\mathcal{H} = \{H \in \mathcal{U}^0 : \limsup \mathrm{Var}\, \psi^{\{H\}}(p_n)/\, \mathrm{Var}\, \psi(G_{n,p_n}) > 0\}. \tag{4.22}$$

Proof. There exists a finite $\mathcal{H}_\varepsilon$ such that (4.20) holds for all large n. Since (4.22) implies that $\mathrm{Var}\, \psi^{\mathcal{H}_\varepsilon \setminus \mathcal{H}}(p_n)/\, \mathrm{Var}\, \psi(G_{n,p_n}) \to 0$, we may replace $\mathcal{H}_\varepsilon$ by $\mathcal{H}_\varepsilon \cap \mathcal{H}$ in (4.20), if we increase ε to 2ε. Hence ψ is almost finitely dominated by $\mathcal{H}$. It follows from the definition (4.18) that every dominating family includes $\mathcal{H}$. $\qquad\square$

REMARK 4.3. If $\limsup p_n < 1$, then the set $\mathcal{H}$ in (4.22) may be written

$$\mathcal{H} = \{H \neq \emptyset : \limsup |\hat{\psi}_n^*(H; p_n)|^2/\, \mathrm{Var}\, \psi(G_{n,p_n}) > 0\}. \tag{4.23}$$

REMARK 4.4. Note that $\mathcal{H}$ is the minimal dominating family when ψ is almost finitely dominated. If ψ is not almost finitely dominated, however, $\mathcal{H}$ may be empty, see the example in Section 11.

We will see in the following sections that also an almost finitely dominated decomposition enables us to deduce limit theorems from Theorem 1.

There remains, however, the case when the decomposition (4.1) is not almost finitely dominated, i.e. when a significant part of the variance comes from terms with larger and larger H as $n \to \infty$. It is then, unfortunately, not possible to use Theorem 1 directly. We may, nevertheless, in some cases obtain limit theorems by the same method as Theorem 1. We also believe that the decomposition (4.1) may be instructive even when it does not immediately yield limit results. We return briefly to this case in the next section.

The following propositions give convenient criteria for almost finite domination.

PROPOSITION 4.7. *Let $p_n \to p < 1$ and suppose that $\mathcal{H}$ is a family of non-empty (unlabelled) graphs and that β_n is a sequence of positive numbers such that*

$$\hat{\psi}_n^*(H; p_n)/\beta_n \to a(H), \qquad H \in \mathcal{H}, \tag{4.24}$$

for some real numbers $a(H)$, not all zero. Then ψ is almost finitely dominated by $\mathcal{H}$ if and only if

$$\mathrm{Var}(\psi(G_{n,p_n}))/\beta_n^2 \to \sum_{\mathcal{H}} a(H)^2 \, \mathrm{aut}(H)(1-p)^{e(H)} < \infty. \tag{4.25}$$

Proof. If $\mathcal{H}'$ is a finite subfamily of $\mathcal{H}$, then by (4.7) and (4.6),

$$\mathrm{Var}(\psi_n^{\mathcal{H}'}(p_n)) = \sum_{H \in \mathcal{H}'} \mathrm{Var}(\hat{\psi}_n(H; p_n) S_n(H; p_n))$$

$$\sim \sum_{H \in \mathcal{H}'} |\hat{\psi}_n^*(H; p_n)|^2 (1-p)^{e(H)} \, \mathrm{aut}(H) \tag{4.26}$$

and thus

$$\mathrm{Var}(\psi_n^{\mathcal{H}'}(p_n))/\beta_n^2 \to \sum_{H \in \mathcal{H}'} a(H)^2 \, \mathrm{aut}(H)(1-p)^{e(H)}. \tag{4.27}$$

Denote the right hand side of (4.27) by $\sum(\mathcal{H}')$. If (4.25) holds, choose $\mathcal{H}_\varepsilon$ such that $\sum(\mathcal{H}_\varepsilon) > (1-\varepsilon)\sum(\mathcal{H})$. Then (4.21) holds for all large n by (4.27) and (4.25).

Conversely, if the decomposition is almost finitely dominated by $\mathcal{H}$, let $\mathcal{H}_\varepsilon$ be as in (4.21). For any finite $\mathcal{H}'$ with $\mathcal{H}_\varepsilon \subseteq \mathcal{H}' \subseteq \mathcal{H}$, (4.27) yields

$$\sum(\mathcal{H}') = \lim \mathrm{Var}(\psi_n^{\mathcal{H}'}(p_n))/\beta_n^2 \leq \liminf \mathrm{Var}(\psi_n^{\mathcal{H}}(p_n))/\beta_n^2$$

$$\leq \limsup \mathrm{Var}(\psi_n^{\mathcal{H}}(p_n))/\beta_n^2 \leq (1-\varepsilon)^{-1} \limsup \mathrm{Var}(\psi_n^{\mathcal{H}_\varepsilon}(p_n))/\beta_n^2$$

$$= (1-\varepsilon)^{-1} \sum(\mathcal{H}_\varepsilon) \leq (1-\varepsilon)^{-1} \sum(\mathcal{H}').$$

Letting $\mathcal{H}' \nearrow \mathcal{H}$ and then $\varepsilon \to 0$ yields (4.25), with $\sum(\mathcal{H}) \leq (1-\varepsilon)^{-1} \sum(\mathcal{H}_\varepsilon) < \infty$. $\qquad \square$

PROPOSITION 4.8. *Let p_n and β_n be sequences of positive numbers with* $\limsup p_n < 1$ *and* $\liminf \operatorname{Var} \psi(G_{n,p_n})/\beta_n^2 > 0$. *Suppose that ψ is dominated by a family $\mathcal{H}$ and that $\mathcal{H}_0$ is a finite subfamily of $\mathcal{H}$ such that*

$$|\hat{\psi}_n^*(H;p_n)| \le \beta_n a(H), \qquad H \in \mathcal{H} \setminus \mathcal{H}_0, \tag{4.28}$$

for all n and some numbers $a(H)$ with

$$\sum_{\mathcal{H} \setminus \mathcal{H}_0} a(H)^2 \operatorname{aut}(H) < \infty. \tag{4.29}$$

Then ψ is almost finitely dominated by $\mathcal{H}$.

Proof. If $\mathcal{H}_0 \subseteq \mathcal{H}_\varepsilon \subseteq \mathcal{H}$, then by (4.2) and (4.5),

$$\limsup_{n \to \infty} \left(\operatorname{Var} \psi^{\mathcal{U}^0 \setminus \mathcal{H}_\varepsilon}(p_n) / \operatorname{Var} \psi(G_n(p_n)) \right)$$

$$= \limsup_{n \to \infty} \left(\operatorname{Var} \psi^{\mathcal{H} \setminus \mathcal{H}_\varepsilon}(p_n) / \operatorname{Var} \psi(G_n(p_n)) \right)$$

$$\le C \limsup_{n \to \infty} \left(\operatorname{Var} \psi^{\mathcal{H} \setminus \mathcal{H}_\varepsilon}(p_n) / \beta_n^2 \right) \le C \sum_{\mathcal{H} \setminus \mathcal{H}_\varepsilon} \operatorname{aut}(H) |\hat{\psi}_n^*(H;p_n)|^2 / \beta_n^2$$

$$\le C \sum_{\mathcal{H} \setminus \mathcal{H}_\varepsilon} \operatorname{aut}(H) a(H)^2,$$

which can be made less than ε by choosing $\mathcal{H}_\varepsilon$ finite but large. $\quad\square$

We have so far studied the decomposition (4.1) of graph statistics, which by definition do not depend on t explicitly. The decomposition may, of course, be extended to more general statistics $\psi(G,t)$ (recall that the projections $\psi^{\mathcal{H}}$ are of this type). The following proposition yields, for example, the decomposition of the drift of $\psi(G_n(t))$.

PROPOSITION 4.9. *If $X(t) = \sum_{H \in \mathcal{U}^0} a(H;t) S_n(H;t)$, where each $a(H;t)$ is continuously differentiable on $[0,1)$, then $X(t)$ is a semimartingale with drift*

$$\xi(t) = \sum_{H \in \mathcal{U}^0} \left(\frac{d}{dt} a(H;t) - \frac{e(H)}{1-t} a(H;t) \right) S_n(H;t), \qquad 0 \le t < 1. \tag{4.30}$$

In particular, $X^{\mathcal{H}}(t)$ has drift $\xi^{\mathcal{H}}(t)$ for every $\mathcal{H} \subseteq \mathcal{U}^0$.

Proof. Itô's lemma (Proposition 2.7) and Lemma 2.1 show that each term $a(H;t) S_n(H;t) = a(H;t)(1-t)^{e(H)} \check{S}_n(H;t)$ is a semimartingale with drift

$$\frac{d}{dt} \left(a(H;t)(1-t)^{e(H)} \right) \check{S}_n(H;t) = \left(\frac{d}{dt} a(H;t) - \frac{e(H)}{1-t} a(H;t) \right) S_n(H;t). \quad\square$$

The drift can also be used to show almost finite domination.

PROPOSITION 4.10. *Let p_n and β_n be sequences of positive numbers with $\limsup p_n < 1$ and $\operatorname{Var} \psi(G_{n,p_n}) \asymp \beta_n^2$. Suppose that $\psi(G_n(t))$ has drift $\xi_n(t)$ and that, for some $A > 1$,*

$$\operatorname{Var} \psi(G_n(t)) = O(\beta_n^2), \qquad p_n \le t \le Ap_n,$$
$$p_n^2 \operatorname{Var} \xi(t) = O(\beta_n^2), \qquad p_n \le t \le Ap_n,$$

uniformly in n and t. Then $\psi(G_{n,p_n})$ is almost finitely dominated by every family that dominates it.

Proof. We may assume that $1 < A < 1/p_n$. Assume that $\mathcal{H}$ dominates ψ, and define $\mathcal{H}_k = \{H \in \mathcal{H} : e(H) \le k\}$, $k \ge 1$. Let $b_n(H;t) = (1-t)^{e(H)} \hat{\psi}_n(H;t)$. Then, $\psi(G_n(t)) = \sum_H b_n(H;t) \check{S}_n(H;t)$, and as in Proposition 4.9, $\xi_n(t) = \sum_H b_n'(H;t) \check{S}_n(H;t)$, and thus

$$\operatorname{Var}(\psi^{\mathcal{H} \backslash \mathcal{H}_k}(G_n(p_n))) = \sum_{\mathcal{H} \backslash \mathcal{H}_k} b(H;p_n)^2 \operatorname{Var} \check{S}_n(H;p_n)$$

$$= \sum_{\mathcal{H} \backslash \mathcal{H}_k} b(H;Ap_n)^2 \operatorname{Var} \check{S}_n(H;p_n)$$

$$- 2 \int_1^A \sum_{\mathcal{H} \backslash \mathcal{H}_k} p_n b(H;sp_n) b'(H;sp_n) \operatorname{Var} \check{S}_n(H;p_n) \, ds$$

$$\le A^{-k} \sum_H b(H;Ap_n)^2 \operatorname{Var} \check{S}_n(H;Ap_n)$$

$$+ 2 \int_1^A \sum_H p_n |b(H;sp_n)| |b'(H;sp_n)| \operatorname{Var} \check{S}_n(H;sp_n) s^{-k} \, ds$$

$$\le A^{-k} \operatorname{Var} \psi(G_n(Ap_n)) + 2 \int_1^A (\operatorname{Var} \psi(G_n(sp_n)))^{1/2} (\operatorname{Var} \xi_n(sp_n))^{1/2} p_n s^{-k} \, ds$$

$$\le C_1 \beta_n^2 \left(A^{-k} + 2 \int_1^A s^{-k} \, ds \right)$$

$$\le C_2 \operatorname{Var} \psi(G_{n,p_n})(A^{-k} + 2/(k-1)). \tag{4.31}$$

Since $A^{-k} + 2/(k-1)$ can be made arbitrarily small by choosing k large, the result follows. $\qquad \square$

Finally, we consider a time reversal as in the proof of Lemma 3.2.

PROPOSITION 4.11. *Let ψ be a graph statistic and define $\overline{\psi}(G) = \psi(-G)$, where $-G$ is the complement of G. Then, for $0 \le t \le 1$,*

$$\overline{\psi}(G_n(t)) \stackrel{d}{=} \psi(G_n(1-t)), \tag{4.32}$$

and

$$\widehat{\overline{\psi}}(H;t) = (-1)^{e(H)} \hat{\psi}(H;1-t). \tag{4.33}$$

Proof. It is obvious that $-G_n(t) \stackrel{d}{=} G_n(1-t)$, which yields (4.32). Furthermore, the natural isomorphism mapping $I_e(t)$ to $1 - I_e(1-t)$ maps $S_n(H;t)$ into

$(-1)^{e(H)}S_n(H; 1 - t)$, cf. the proof of Lemma 3.2. Hence, by (4.4), (provided $0 < t < 1$; the cases $t = 0, 1$ follow by continuity),

$$\widehat{\overline{\psi}}(H; t) = \mathrm{E}(\psi(-G_n(t))S_n(H; t))/\mathrm{E}\, S_n(H; t)^2$$
$$= \mathrm{E}\big(\psi(G_n(1 - t))(-1)^{e(H)}S_n(H; 1 - t)\big)/\mathrm{E}\, S_n(H; 1 - t)^2$$
$$= (-1)^{e(H)}\hat{\psi}(H; 1 - t). \qquad \square$$

II. LIMIT THEOREMS

5. Limits for $G_{n,p}$

In this section we suppose that p_n is a given sequence in $(0,1)$, and that ψ is a graph statistic. We are interested in the asymptotics of the distribution of $\psi(G_{n,p_n}) = \psi(G_n(p_n))$. We will often assume that $p = \lim p_n$ exists; this is no real loss of generality, since Proposition 2.5 implies that if the obtained limit does not depend on p, then any sequence p_n gives the same limit. For simplicity, we will also often ignore the case $p = 1$, which can be treated by replacing the graphs by their complements and p_n by $1 - p_n$, cf. Proposition 4.11. We will also ignore to mention the trivial exceptions that may arise when $\psi(G_{n,p_n})$ is constant.

Suppose first that ψ has a finite decomposition. Then the asymptotics follow directly from this decomposition and Theorem 1 (choosing $t = 1$ in (3.2)); the form of the limit (if it exists) and the appropriate normalization are determined by the asymptotics of the coefficients in the decomposition. This generalizes easily to graph statistics with finitely dominated decompositions, and further to almost finitely dominate decompositions, so we state the result in this generality.

THEOREM 2. *Suppose that $p_n \to p \geq 0$, that β_n is a sequence of positive numbers, and that ψ is almost finitely dominated by a family $\mathcal{H}$ of non-empty graphs such that, for every $H \in \mathcal{H}$,*

$$np_n^{m(H)} \to \infty \tag{5.1}$$

and

$$a(H) = \lim_{n\to\infty} \hat{\psi}_n^*(H; p_n)/\beta_n \quad \text{exists.} \tag{5.2}$$

Then, with $X_n = \psi(G_{n,p_n})$,

$$\frac{X_n - \mathrm{E}\, X_n}{\beta_n} \xrightarrow{d} \sum_{H \in \mathcal{H}} a(H)U(H; 1) \tag{5.3}$$

where $U(H; 1)$ are as in Theorem 1. (If $\mathcal{H}$ is infinite, the sum is interpreted in L^2.) Furthermore,

$$\mathrm{Var}(X_n)/\beta_n^2 \to \sum_{H \in \mathcal{H}} a(H)^2 \operatorname{aut}(H)(1 - p)^{e(H)} < \infty. \tag{5.4}$$

Proof. Proposition 4.7 yields (5.4) (the cases when all $a(H)$ vanish or $p = 1$ are easy).

Let $\mathcal{H}_\varepsilon$ be as in (4.20) and define, for $\nu, n \geq 1$, the projection

$$Z_{\nu n} = \beta_n^{-1} \psi_n^{\mathcal{H}_{1/\nu}}(p_n) = \sum_{H \in \mathcal{H}_{1/\nu}} \beta_n^{-1} \hat{\psi}_n^*(H; p_n) n^{-v(H)/2} p_n^{-e(H)/2} S_n(H; p_n) \quad (5.5)$$

and

$$Z_\nu = \sum_{H \in \mathcal{H}_{1/\nu}} a(H) U(H; 1). \tag{5.6}$$

By Theorem 1, $Z_{\nu n} \xrightarrow{d} Z_\nu$ as $n \to \infty$ for every ν.

We may assume that $\mathcal{H}_{1/\nu} \nearrow \mathcal{H}$, since we may otherwise increase $\mathcal{H}_{1/\nu}$. By Theorem 1 and (5.4),

$$\sum_{\mathcal{H}} a(H)^2 \operatorname{Var} U(H; 1) = \sum_{\mathcal{H}} a(H)^2 \operatorname{aut}(H)(1 - p)^{e(H)} < \infty, \tag{5.7}$$

and consequently the orthogonality of $\{U(H; 1)\}$ yields

$$Z_\nu \to \sum_{H \in \mathcal{H}} a(H) U(H; 1) \qquad \text{as } \nu \to \infty \tag{5.8}$$

in L^2, and thus in distribution.

Finally, the assumption that $\mathcal{H}$ dominates almost finitely yields

$$\limsup_{n \to \infty} \operatorname{E} |\beta_n^{-1}(X_n - \operatorname{E} X_n) - Z_{\nu n}|^2 \leq \frac{1}{\nu} \lim_{n \to \infty} \operatorname{Var}(X_n)/\beta_n^2$$

$$\to 0 \qquad \text{as } \nu \to \infty, \tag{5.9}$$

and the result follows by [4, Theorem 4.2]. $\qquad \square$

The simplest case is when every $H \in \mathcal{H}$ is connected; then $U(H; 1)$, $H \in \mathcal{H}$, are independent and normal, and the theorem thus yields a normal limit. In this case we can relax and simplify the assumptions.

THEOREM 3. Let $X_n = \psi(G_{n,p_n})$. Suppose that $\limsup p_n < 1$ and that ψ is almost finitely dominated by a family $\mathcal{H}$ of non-empty connected graphs such that

$$np_n^{m(H)} \to \infty \qquad \text{for every } H \in \mathcal{H}. \tag{5.10}$$

Then

$$(\operatorname{Var} X_n)^{-\frac{1}{2}}(X_n - \operatorname{E} X_n) \xrightarrow{d} N(0, 1). \tag{5.11}$$

Proof. Let $\sigma_n^2 = \operatorname{Var}(X_n)$. Then, by (4.6), for $H \in \mathcal{H}$,

$$\sigma_n^2 \geq \operatorname{Var}(\hat{\psi}_n(H; p_n) S_n(H; p_n))$$

$$= |\hat{\psi}_n^*(H; p_n)|^2 (1 + O(\tfrac{1}{n}))(1 - p_n)^{e(H)} \operatorname{aut}(H) \tag{5.12}$$

and thus $\limsup_{n \to \infty} |\hat{\psi}_n^*(H; p_n)|^2/\sigma_n^2 \leq \limsup(1 - p_n)^{-e(H)} < \infty$.

Hence the sequence $a_n(H) = \hat{\psi}_n^*(H; p_n)/\sigma_n$ is bounded for every $H \in \mathcal{H}$. For any subsequence of the natural numbers, we can thus select a subsubsequence such that $a_n(H)$ converges for a given H. By the diagonal method there exists a subsubsequence such that $a_n(H)$ converges for every $H \in \mathcal{H}$; we may further assume that also p_n converges. Theorem 2 (for subsequences) shows that along this subsubsequence

$$\sigma_n^{-1}(X_n - \mathrm{E}\, X_n) \xrightarrow{d} \sum_{H \in \mathcal{H}} a(H) U(H; 1) \sim N(0, \sigma^2),$$

with $a(H) = \lim a_n(H)$ and

$$\sigma^2 = \sum_{H \in \mathcal{H}} a(H)^2 \operatorname{Var} U(H; 1) = \lim_{n \to \infty} \operatorname{Var}(X_n)/\sigma_n^2 = 1.$$

Hence every subsequence of $\sigma_n^{-1}(X_n - \mathrm{E}\, X_n)$ has a subsubsequence which converges to $N(0, 1)$, which by Proposition 2.5 implies that the whole sequence converges. $\qquad\square$

REMARK 5.1. Theorem 2 holds with joint convergence in (5.3) for any (finite) set of statistics ψ that satisfy the conditions (possibly with different β_n). This follows from the proof above, or by the Cramér–Wold device. In particular, if the statistics are dominated by families of graphs without common components (but typically not otherwise), the limits are independent.

Similarly, if several ψ satisfy the assumptions of Theorem 3, we obtain convergence to a joint normal distribution whose covariance matrix is the limit of the correlation matrix of $\psi(G_{n,p_n})$, provided this limit exists.

If ψ has a finitely dominated decomposition (in particular, if the decomposition is finite), the condition in Theorem 3 that it be dominated by a family of connected graphs is also necessary for asymptotic normality. We need an elementary lemma.

LEMMA 5.1. *Suppose that q is a polynomial in m variables of degree ≥ 2, and that $\xi_1, \ldots, \xi_m$ are independent non-degenerate normal variables. Then $q(\xi_1, \ldots, \xi_m)$ is not normally distributed.*

Proof. We may assume that $\xi_1, .., \xi_m \sim N(0, 1)$; we may also assume that $m \geq 2$, since we otherwise may add an extra variable. Let d be the degree of q and write $\xi = (\xi_1, \ldots, \xi_m)$, $\xi' = \xi/|\xi|$. There exist $\xi_0' \in \mathbf{R}^m$ with $|\xi_0'| = 1$, $\varepsilon > 0$ and $\delta > 0$, such that if $|\xi' - \xi_0'| < \varepsilon$ and $|\xi|$ is large enough, then $|q(\xi)| \geq \delta|\xi|^d$. Consequently, for large t,

$$\mathrm{P}(|q(\xi)| > t) \geq \mathrm{P}(|\xi| \geq (t/\delta)^{1/d}, \, |\xi' - \xi_0'| < \varepsilon)$$
$$= c \int_{(t/\delta)^{1/d}}^{\infty} r^{m-1} e^{-r^2/2} dr$$
$$\geq c \int_{(t/\delta)^{1/d}}^{\infty} r e^{-r^2/2} dr = c e^{-c_1 t^{2/d}} \qquad (5.13)$$

with $c, c_1 > 0$, and thus $q(\xi)$ is not normal. $\qquad\square$

THEOREM 4. *Let $X_n = \psi(G_{n,p_n})$. Suppose that $\limsup p_n < 1$ and that ψ is dominated by a finite family $\mathcal{H}$ such that (5.10) holds. If $\mathcal{H}$ is a minimal dominating family and contains a disconnected graph, then there are no sequences α_n and β_n such that $(X_n - \alpha_n)/\beta_n \xrightarrow{d} N(0,1)$.*

Proof. Define σ_n and $a_n(H)$ as in the proof of Theorem 3. It follows again that (5.12) holds and that each sequence $a_n(H)$ is bounded. Furthermore, $a_n(H)$ does not converge to 0 for any $H \in \mathcal{H}$, because that would imply that $\mathcal{H} \setminus \{H\}$ is another dominating family and hence $\mathcal{H}$ would not be minimal. Let H_0 be a disconnected graph in $\mathcal{H}$. It follows that there exists a subsequence of the natural numbers such that $a_n(H_0)$ converges to a non-zero limit $a(H_0)$ along this subsequence. We may further assume, as in the proof of Theorem 3, that $a_n(H)$, for every $H \in \mathcal{H}$, and p_n converge along this subsequence. By Theorem 2, along the subsequence,

$$\sigma_n^{-1}(X_n - \mathrm{E}\,X_n) \xrightarrow{d} Z = \sum a(H)U(H;1).$$

By Theorem 1, Z is a polynomial in the independent normal variables $\{U(H';1) : H' \text{ equals a component of some } H \in \mathcal{H}\}$, and the degree is at least 2 because $a(H_0) > 0$. Consequently, by Lemma 5.1, Z is not normal.

If also $\beta_n^{-1}(X_n - \alpha_n) \xrightarrow{d} W$, for some non-degenerate W, along the same subsequence, then Z equals in distribution a linear function of W; in particular, $\beta_n^{-1}(X_n - \alpha_n) \xrightarrow{d} N(0,1)$ cannot hold. $\square$

We can combine Theorems 3 and 4 into the following criterion.

THEOREM 5. *Let $X_n = \psi(G_{n,p_n})$. Suppose that $\limsup p_n < 1$, and that ψ is dominated by a finite family $\mathcal{H}$ such that (5.10) holds. Then X_n is asymptotically normal if and only if $\hat{\psi}_n^*(H;p_n)/(\mathrm{Var}(X_n))^{1/2} \to 0$ for every disconnected $H \in \mathcal{H}$.* $\square$

In Theorems 4 and 5 we need $\mathcal{H}$ to be finite; the results are not valid if we only assume that ψ is almost finitely dominated by $\mathcal{H}$, as is shown by the following example. The example, however, is rather artificial, and it seems reasonable to regard the presence of disconnected graphs in a minimal dominating family as a strong indication that the variables are not asymptotically normal.

EXAMPLE 5.1. Let $\xi \sim N(0,1)$ and define f by $f(x) = \begin{cases} x, & |x| \geq A, \\ -x, & |x| < A, \end{cases}$ where $A > 0$ is chosen such that $\mathrm{E}(\xi f(\xi)) = 0$. Since the Hermite polynomials h_k form an orthogonal basis in $L^2((2\pi)^{-\frac{1}{2}}e^{-x^2/2}dx)$, $f(\xi) = \sum_0^\infty a_k h_k(\xi) = \sum_0^\infty a_k{:}\xi^k{:}$ for some a_k, with $a_0 = \mathrm{E}\,f(\xi) = 0$ and $a_1 = \mathrm{E}(f(\xi)\xi) = 0$. Let $p_n = 1/2$, $\psi(G) = f(\sqrt{8}(e(G) - \frac{1}{2}\binom{v(G)}{2})/v(G))$ and $X_n = \psi(G_{n,1/2}) = f(n^{-1}2^{1/2}S_n(K_2;1/2))$.

Let $U_n^{(k)} = n^{-k}2^{k/2}S_n(kK_2;\frac{1}{2})$. Theorem 1 yields, for $k \geq 1$,

$$(U_n^{(1)}, U_n^{(k)}) \xrightarrow{d} (U(K_2;1), {:}U(K_2;1)^k{:}) \overset{d}{=} (\xi, h_k(\xi)),$$

and thus, since f is measurable and a.s. continuous at ξ,

$$(X_n, U_n^{(k)}) \xrightarrow{d} (f(\xi), h_k(\xi)).$$

Since the variables are uniformly square integrable, this implies

$$\hat{\psi}_n^*(kK_2; \tfrac{1}{2}) = n^k(\tfrac{1}{2})^{k/2}\hat{\psi}(kK_2; \tfrac{1}{2}) = \mathrm{E}(X_n U_n^{(k)})/\mathrm{E}(U_n^{(k)})^2$$
$$\to \mathrm{E}(f(\xi)h_k(\xi))/\mathrm{E}(h_k(\xi))^2 = a_k.$$

Since further $\operatorname{Var} X_n \to \operatorname{Var} f(\xi) = \sum_2^\infty a_k^2 \mathrm{E}(h_k(\xi))^2 = \sum_2^\infty a_k^2 k!$, Proposition 4.7 shows that X_n is almost finitely dominated by $\{kK_2\}_2^\infty$. These graphs are all disconnected; nevertheless $X_n \xrightarrow{d} f(\xi) \overset{d}{=} \xi \sim N(0,1)$.

If ψ is not even almost finitely dominated, the decomposition does not directly yield limit results. In such cases it may be possible to use a martingale convergence theorem directly, for example as follows.

THEOREM 6. *Let* $X_n = \psi(G_{n,p_n})$ *for some* $p_n > 0$ *and let* $\sigma_n^2 = \operatorname{Var}(X_n)$. *Suppose that there exists, for each* $n \geq 1$, *a martingale* $M_n(t)$ *and a continuous function* $\beta_n^2(t)$ *on* $[0, p_n]$ *such that* $M_n(0) = 0$ *and*

$$\sup_{0 \leq t \leq p_n} \mathrm{E}\,|[M_n, M_n]_t - \beta_n^2(t)| = o(\sigma_n^2), \tag{5.14}$$
$$\operatorname{Var}\big(\psi(G_n(p_n)) - M_n(p_n)\big) = o(\sigma_n^2). \tag{5.15}$$

Then

$$\sigma_n^{-1}(X_n - \mathrm{E}\,X_n) \xrightarrow{d} N(0,1). \tag{5.16}$$

Proof. $\mathrm{E}[M_n, M_n]_{p_n} = \operatorname{Var}(M_n(p_n)) \sim \sigma_n^2$ and hence $\beta_n^2(p_n) \sim \sigma_n^2$. [8, Theorem VIII.3.12] and a change of time yields

$$\sigma_n^{-1} M_n(p_n) \xrightarrow{d} N(0,1),$$

see [2, Lemma 8.2] for details, and the result follows by (5.15).   □

It remains to find a suitable M_n. One possibility is to define

$$M_n(t) = \sum_{H \neq \emptyset} \hat{\psi}_n(H; p_n)(1 - p_n)^{e(H)} \check{S}_n(H; t), \tag{5.17}$$

which is a martingale with $M_n(p_n) = \psi_n(G_n(p_n)) - \mathrm{E}\,\psi_n(G_n(p_n))$. It may, however, be difficult to verify (5.14) for this choice and it may be better to use some other M_n found by trial and error. One such example is treated in [2].

6. Limits for $G_{n,m}$

Let m_n be a given sequence of integers with $0 \le m_n \le \binom{n}{2}$. We will assume that $m_n \to \infty$, and will study the asymptotics of $\psi(G_{n,m_n})$, where ψ is a graph statistic.

We begin by defining $p_n = m_n/\binom{n}{2}$, the density of edges in G_{n,m_n}. A general rule in random graph theory is that G_{n,m_n} and G_{n,p_n} usually behave similarly. We will show results of this type in the present section, but there are important qualifications and the asymptotics for G_{n,m_n} and G_{n,p_n} are sometimes quite different. This should not be surprising, as the following example shows.

EXAMPLE 6.1. Let us replace ψ by $\tilde{\psi}(G) = \psi(G) + a_n(e(G) - m_n)$ for some constants a_n, where $n = v(G)$; then $\tilde{\psi}(G_{n,m_n}) = \psi(G_{n,m_n})$. Further, $\hat{\tilde{\psi}}_n(H;t) = \hat{\psi}_n(H;t)$ for all H except $\emptyset$ and K_2, but $\hat{\tilde{\psi}}_n(K_2;t) = \hat{\psi}_n(K_2;t) + \frac{1}{2}a_n$; in particular, the choice $a_n = -2\hat{\psi}_n(K_2;p_n)$ makes $\hat{\tilde{\psi}}_n(K_2;p_n) = 0$. This shows that the asymptotic distribution of $\psi(G_{n,m_n})$ cannot depend on $\hat{\psi}_n(K_2;p_n)$. Furthermore, if $\psi(G_{n,p_n})$ is dominated by K_2 (i.e., the method of the first projection is applicable, see Example 4.1), $\tilde{\psi}(G_{n,p_n})$ has variance of a smaller order of magnitude than $\psi(G_{n,p_n})$, and hence the asymptotic variances for G_{n,m_n} and G_{n,p_n} differ for at least one of $\psi, \tilde{\psi}$.

We define $\tau_n = \min\{t : e(G_n(t)) \ge m_n\}$, i.e. the m_nth order statistic of $T_1, \dots, T_{\binom{n}{2}}$. Then $G_n(\tau_n)$ has a.s. m_n edges and has the distribution of G_{n,m_n}. One reason for the similarity between $G_{n,m_n} = G_n(\tau_n)$ and $G_{n,p_n} = G_n(p_n)$ is that τ_n is close to p_n; more precisely, we have the following elementary (and well-known) results.

LEMMA 6.1. If $m_n \to \infty$, then

$$\tau_n/p_n \xrightarrow{p} 1. \tag{6.1}$$

More precisely,

$$\tau_n = p_n - (n(n-1))^{-1} S_n(K_2, \tau_n) \tag{6.2}$$

and, if $p_n \to p \ge 0$,

$$np_n^{-1/2}(\tau_n - p_n) \xrightarrow{d} N(0, 2(1-p)). \tag{6.3}$$

Proof. (6.1) is just a version of the weak law of large numbers; in fact, Chebyshev's inequality yields

$$\mathrm{P}(\tau_n \le (1-\varepsilon)p_n) = \mathrm{P}(\mathrm{Bi}(\binom{n}{2}, (1-\varepsilon)p_n) \ge \binom{n}{2}p_n)$$
$$\le (\varepsilon\binom{n}{2}p_n)^{-2}\binom{n}{2}(1-\varepsilon)p_n < (m_n\varepsilon^2)^{-1} \to 0,$$

and similarly $\mathrm{P}(\tau_n > (1+\varepsilon)p_n) \to 0$, for every $\varepsilon > 0$. By the definition of $S_n(K_2)$,

$$S_n(K_2, \tau_n) = 2(e(G_n(\tau_n)) - \binom{n}{2}\tau_n) = n(n-1)(p_n - \tau_n), \tag{6.4}$$

which yields (6.2). Finally, it follows from (6.1), Proposition 2.1 and Theorem 1, writing $\tau_n = p_n(\tau_n/p_n)$, that

$$n^{-1}p_n^{-1/2}S_n(K_2;\tau_n) \xrightarrow{d} U(K_2;1) \subset N(0,2(1-p)), \qquad (6.5)$$

which yields (6.3). $\qquad\qquad\qquad\qquad\qquad\qquad\qquad\qquad\qquad\qquad\square$

LEMMA 6.2. *Assume that* $m_n \to \infty$. *If* Δ_n *is any sequence of random variables such that* $\Delta_n I(|\tau_n - p_n| \le \lambda p_n^{1/2}/n) \xrightarrow{p} 0$ *as* $n \to \infty$, *for every fixed* $\lambda > 0$, *then* $\Delta_n \xrightarrow{p} 0$.

Proof. Assume that $p_n \to p \le 1$; the general case follows by considering subsequences. Let $\varepsilon > 0$ and $\lambda > 0$. By assumption and Lemma 6.1, with a minor modification if $p = 1$,

$$\limsup_{n\to\infty} P(|\Delta_n| > \varepsilon) \le \lim_{n\to\infty} P(|\tau_n - p_n| > \lambda p_n^{1/2}/n) = 2(1 - \Phi(\lambda/\sqrt{2(1-p)})),$$

and the result follows by letting $\lambda \to \infty$. $\qquad\qquad\qquad\qquad\qquad\square$

We will use the decomposition of Section 4, which now takes the form

$$\psi(G_{n,m_n}) = \psi(G_n(\tau_n)) = \sum_H \hat{\psi}_n(H;\tau_n)S_n(H;\tau_n). \qquad (6.6)$$

(Since τ_n is random, the terms in this sum are in general not orthogonal.) Theorem 1, (6.1) and Proposition 2.1 yield, just as for (6.5),

$$n^{-v(H)/2}p_n^{-e(H)/2}S_n(H;\tau_n) \xrightarrow{d} U(H;1), \qquad H \in \mathcal{U}^0 \qquad (6.7)$$

provided $np_n^{m(H)} \to \infty$, jointly in any finite set of such H, but, since the coefficients in (6.6) are random, we do not immediately obtain a limit for $\psi(G_{n,m_n})$. In particular, we want to normalize $\psi(G_{n,m_n})$ by subtracting $E\psi(G_{n,m_n})$, or another suitable constant, and not the random $\hat{\psi}_n(\emptyset;\tau_n)$. It turns out that this leads to important cancellations in (6.6), as suggested by Example 6.1.

We begin with the case of a finite decomposition.

THEOREM 7. *Let* $X_n = \psi(G_{n,m_n})$, *where* ψ *has a finite decomposition and* $m_n \to \infty$. *Let* $p_n = m_n/\binom{n}{2}$ *and suppose that* $p_n \to p < 1$. *Suppose further that* β_n *is a sequence of positive numbers such that for every* $H \in \mathcal{U}^0$ *with at least three vertices, there exists* $a(H)$ *such that*

$$\hat{\psi}_n^*(H;p_n)/\beta_n \to a(H), \qquad (6.8)$$

and, further, if H *is such a graph and* $a(H) \ne 0$, *then* $np_n^{m(H)} \to \infty$. *Then*

$$\frac{X_n - \alpha_n}{\beta_n} \xrightarrow{d} \sum_{H \in \tilde{\mathcal{U}}} \tilde{a}(H)U(H;1), \qquad (6.9)$$

where $\widetilde{\mathcal{U}}$ is the set of all nonempty graphs where every component has at least three vertices, $U(H;1)$ is as in Theorem 1,

$$\alpha_n = \hat{\psi}_n(\emptyset; p_n) + \beta_n \sum_{j \geq 1} (-1)^j \frac{(2j)!}{j!} (1-p)^j a(2jK_2), \tag{6.10}$$

and

$$\tilde{a}(H) = \sum_{j \geq 0} (-1)^j \frac{(2j)!}{j!} (1-p)^j a(H + 2jK_2). \tag{6.11}$$

In particular, if $a(H) = 0$ for every $H \in \mathcal{U}^0$ that has two or more components with three or more vertices, but $\tilde{a}(H) \neq 0$ for some $H \in \widetilde{\mathcal{U}}$, then $\psi(G_{n,m_n})$ is asymptotically normal.

REMARK 6.1. If $a(H) = 0$ for every H that has at least two components K_2, we thus obtain the same limit as for G_{n,p_n} in Theorem 2, except that only the terms for H with no component K_2 are used.

Proof. We assume for simplicity that $\beta_n = 1$; the general case follows by replacing $\psi(G)$ by $\psi(G)/\beta_{v(G)}$. Then, (6.8) implies

$$\hat{\psi}_n(H; p_n) = O(n^{-v(H)/2} p_n^{-e(H)/2}), \tag{6.12}$$

when $v(H) \geq 3$. Proposition 4.3 and induction shows that, for every H and $k \geq 0$,

$$\frac{d^k}{dt^k} \hat{\psi}_n(H; t) = \sum c(H, H') \hat{\psi}_n(H'; t)(n - v(H))_{v(H')-v(H)}, \tag{6.13}$$

where the sum is taken over graphs H' with $e(H') = e(H) + k$ and $v(H') \leq v(H) + 2k$, and $c(H, H')$ are some combinatorial coefficients; the only H' in this sum with $v(H') = v(H) + 2k$ is $H + kK_2$, and

$$c(H, H + kK_2) = 2^{-k} \operatorname{aut}(H + kK_2) / \operatorname{aut}(H). \tag{6.14}$$

It follows, using (6.12) and treating the cases $k = 0$ and $k = 1$, $H = \emptyset$ separately, that

$$\frac{d^k}{dt^k} \hat{\psi}_n(H; p_n) = 2^{-k} \frac{\operatorname{aut}(H + kK_2)}{\operatorname{aut}(H)} (n - v(H))_{2k} \, \hat{\psi}_n(H + kK_2; p_n)$$
$$+ O\Big(\sum_{H' \neq H + kK_2} n^{-v(H')/2} p_n^{-e(H)/2 - k/2} n^{v(H')-v(H)} \Big)$$
$$= 2^{-k} \frac{\operatorname{aut}(H + kK_2)}{\operatorname{aut}(H)} (n(n-1))^k \hat{\psi}_n(H + kK_2; p_n) + o(n^{-v(H)/2+k} p_n^{-e(H)/2 - k/2}). \tag{6.15}$$

In particular, if $v(H) \geq 3$ or $k \geq 2$, by (6.12),

$$\frac{d^k}{dt^k} \hat{\psi}_n(H; p_n) = O(n^{-v(H)/2+k} p_n^{-e(H)/2 - k/2}). \tag{6.16}$$

Since the decomposition is finite, (6.13) implies that $\frac{d^k}{dt^k}\hat{\psi}_n(H;t)$ vanishes if k is large so, as observed in Proposition 4.5, every $\hat{\psi}_n(H)$ is a polynomial in t with uniformly bounded degree. We can therefore use a finite Taylor expansion and obtain by (6.16), if t_n is any sequence in $(0,1)$ with $t_n = p_n + O(p_n^{1/2}/n)$, and $v(H) \geq 3$,

$$\hat{\psi}_n(H;t_n) = \sum_{k \geq 0} \frac{1}{k!}\frac{d^k}{dt^k}\hat{\psi}_n(H;p_n)(t_n - p_n)^k = O(n^{-v(H)/2}p_n^{-e(H)/2}). \quad (6.17)$$

Moreover, $t_n - p_n = O(p_n^{1/2}/n) = O(p_n/m_n^{1/2}) = o(p_n)$ and thus $t_n \sim p_n$; hence we have proved that if (6.12) holds for a sequence p_n (and every H with $v(H) \geq 3$), then it holds with p_n replaced by any sequence $t_n = p_n + O(p_n^{1/2}/n)$. The argument above then shows that also (6.15) holds for any such sequence t_n. It follows by Lemma 6.2 that, for every $H \in \mathcal{U}^0$,

$$n^{v(H)/2-k}p_n^{e(H)/2+k/2}\left(\frac{d^k}{dt^k}\hat{\psi}_n(H;\tau_n)\right.$$
$$\left. - 2^{-k}\frac{\text{aut}(H + kK_2)}{\text{aut } H}(n(n-1))^k\hat{\psi}_n(H + kK_2;\tau_n)\right)$$
$$\xrightarrow{p} 0; \quad (6.18)$$

in fact, if Δ_n denotes the left hand side of (6.18), and $\lambda > 0$ is fixed, we have just shown that $\Delta_n I(|\tau_n - p_n| \leq \lambda p_n^{1/2}/n) \to 0$ a.s., and thus in probability.

Now suppose that no component of H is a K_1 or K_2, and let $\sigma_n^2 = 2n^2p_n(1-p)$. We claim that, for every $l \geq 0$,

$$\hat{\psi}_n(H+lK_2;\tau_n)\left(S_n(H+lK_2;\tau_n) - \sum_{j=0}^{[l/2]}\frac{(-1)^j l!}{2^j j!(l-2j)!}\sigma_n^{2j}S_n(H;\tau_n)S_n(K_2;\tau_n)^{l-2j}\right)$$
$$\xrightarrow{p} 0. \quad (6.19)$$

We prove this by considering three cases separately, denoting the left hand side of (6.19) by Δ'_n.

First, if $l = 0$ or if $H = \emptyset$ and $l = 1$, then Δ'_n vanishes identically. In the remaining cases, we may thus assume $l \geq 1$ and $v(H + lK_2) \geq 4$; in particular we may use (6.8).

If $a(H + rK_2) \neq 0$ for some $r \geq l$, then, by assumption, $np_n^{m(H+rK_2)} \to \infty$. Since $m(H + rK_2) = m(H + lK_2) = m(H) \vee m(K_2)$, we can use Theorem 1 for $H + lK_2, H, K_2$ and obtain, cf. (6.5),

$$n^{-v(H)/2-l}p_n^{-e(H)/2-l/2}\left(S_n(H + lK_2;\tau_n)\right.$$
$$\left. - \sum_{j}\frac{(-1)^j l!}{2^j j!(l-2j)!}\sigma_n^{2j}S_n(H;\tau_n)S_n(K_2;\tau_n)^{l-2j}\right)$$
$$\xrightarrow{d} U(H + lK_2;1) - \sum_{j}\frac{(-1)^j l!}{2^j j!(l-2j)!}U(H;1)U(K_2;1)^{l-2j}(2(1-p))^j$$
$$= 0, \quad (6.20)$$

using (3.4), and $E U(K_2,1)^2 = 2(1-p)$. For any $\lambda > 0$, (6.17) implies that
$\hat{\psi}_n(H + lK_2; \tau_n) \cdot I(|\tau_n - p_n| \leq \lambda p_n^{1/2}/n) = O(n^{-v(H)/2-l}p_n^{-e(H)/2-l/2})$, which
together with (6.20) yields

$$\Delta'_n I(|\tau_n - p_n| \leq \lambda p_n^{1/2}/n) \xrightarrow{p} 0; \tag{6.21}$$

(6.19) now follows by Lemma 6.2.

Finally, if $a(H + rK_2) = 0$ for every $r \geq l$, then $\hat{\psi}_n^*(H + (k+l)K_2; p_n) \to 0$
and (6.15) yields

$$\frac{d^k}{dt^k}\hat{\psi}_n(H + lK_2; p_n) = o(n^{-v(H)/2-l+k}p_n^{-e(H)/2-l/2-k/2}), \tag{6.22}$$

for every $k \geq 0$. Thus, by Taylor expansion as in (6.17),

$$\hat{\psi}_n(H + lK_2; \tau_n)I(|\tau_n - p_n| \leq \lambda p_n^{1/2}/n) = o(n^{-v(H)/2-l}p_n^{-e(H)/2-l/2}), \tag{6.23}$$

for every $\lambda > 0$. Moreover, Lemma 3.2 implies that if $\lambda > 0$ is fixed and n is
large,

$$E(|S_n(H; \tau_n)|^2 I(|\tau_n - p_n| \leq \lambda p_n^{1/2}/n)) \leq E(\sup_{t \leq 2p_n} |S_n(H; t)|)^2$$
$$= O(n^{v(H)}p_n^{e(H)}), \tag{6.24}$$

for any H without isolated vertices. Since $|\tau_n - p_n| \leq \lambda p_n^{1/2}/n$ implies, using
(6.4), $\sigma_n^{2j}S_n(K_2; \tau_n)^{l-2j} = O((np_n^{1/2})^l)$, we obtain from (6.23), (6.24) and the
corresponding estimate for $H + lK_2$,

$$E(\Delta'_n I(|\tau_n - p_n| \leq \lambda p_n^{1/2}/n))^2 \to 0, \tag{6.25}$$

and thus (6.19) follows as before by Lemma 6.2. This proves (6.19) for all cases.

We now sum (6.19) over l, recalling that all sums below really are finite (with
a fixed number of terms independent of n) and obtain, using $Y_n \approx Z_n$ to denote
$Y_n - Z_n \xrightarrow{p} 0$,

$$\sum_{l \geq 0} \hat{\psi}_n(H + lK_2; \tau_n)S_n(H + lK_2; \tau_n)$$

$$\approx \sum_{j,k \geq 0} \frac{(-1)^j(2j+k)!}{2^j j! k!}\sigma_n^{2j}\hat{\psi}(H + (2j+k)K_2; \tau_n)S_n(H; \tau_n)S_n(K_2; \tau_n)^k$$

$$= \sum_{j \geq 0}(-1)^j\frac{(2j)!}{2^j j!}\sigma_n^{2j}\sum_{k \geq 0}\frac{1}{k!}\frac{(2j+k)!}{(2j)!}\hat{\psi}_n(H + (2j+k)K_2; \tau_n)$$
$$\cdot\left(n(n-1)(p_n - \tau_n)\right)^k S_n(H; \tau_n)$$

$$\approx \sum_{j \geq 0}(-1)^j\frac{(2j)!}{2^j j!}\sigma_n^{2j}\sum_{k \geq 0}\frac{1}{k!}\frac{d^k}{dt^k}\hat{\psi}_n(H + 2jK_2, \tau_n)(p_n - \tau_n)^k S_n(H; \tau_n)$$

$$= \sum_{j \geq 0}(-1)^j\frac{(2j)!}{2^j j!}\sigma_n^{2j}\hat{\psi}_n(H + 2jK_2; p_n)S_n(H; \tau_n)$$

$$= \sum_{j \geq 0}(-1)^j\frac{(2j)!}{j!}(1-p)^j\hat{\psi}_n^*(H + 2jK_2; p_n)n^{-v(H)/2}p_n^{-e(H)/2}S_n(H; \tau_n), \tag{6.26}$$

where the second approximation follows by Lemma 6.2 from (6.18) applied to $H + 2jK_2$ and (6.24), because (when no component of H is K_2)

$$2^{-k}\frac{\text{aut}(H + (2j + k)K_2)}{\text{aut}(H + 2jK_2)} = 2^{-k}\frac{2^{2j+k}(2j + k)! \text{ aut } H}{2^{2j}(2j)! \text{ aut } H} = \frac{(2j + k)!}{(2j)!}. \qquad (6.27)$$

Write the right hand side of (6.26) as $\tilde{a}_n(H)n^{-v(H)/2}p_n^{-e(H)/2}S_n(H; \tau_n)$. Then, by (6.8),

$$\tilde{a}_n(\emptyset) \approx \hat{\psi}_n(\emptyset; p_n) + \sum_{j\geq 1}(-1)^j\frac{(2j)!}{j!}(1 - p)^j a(2jK_2) = \alpha_n, \qquad (6.28)$$

while, if $H \in \widetilde{\mathcal{U}}$,

$$\tilde{a}_n(H) \to \sum_{j\geq 0}(-1)^j\frac{(2j)!}{j!}(1 - p)^j a(H + 2jK_2) = \tilde{a}(H). \qquad (6.29)$$

Consequently, summing (6.26) over $\{\emptyset\} \cup \widetilde{\mathcal{U}}$, and using (6.24) and Lemma 6.2,

$$\begin{aligned}
\psi(G_n(\tau_n)) &= \sum_{H\in\{\emptyset\}\cup\widetilde{\mathcal{U}}}\sum_{l\geq 0}\hat{\psi}_n(H + lK_2; \tau_n)S_n(H + lK_2; \tau_n) \\
&\approx \sum_{H\in\{\emptyset\}\cup\widetilde{\mathcal{U}}}\tilde{a}_n(H)n^{-v(H)/2}p_n^{-e(H)/2}S_n(H; \tau_n) \\
&\approx \alpha_n + \sum_{H\in\widetilde{\mathcal{U}}}\tilde{a}_n(H)n^{-v(H)/2}p_n^{-e(H)/2}S_n(H; \tau_n). \qquad (6.30)
\end{aligned}$$

By (6.24), we may here ignore every term with $\tilde{a}_n(H) \to 0$. In the remaining cases, $\tilde{a}(H) \neq 0$ and thus $a(H + 2jK_2) \neq 0$ for some $j \geq 0$, and the assumptions yield $np_n^{m(H)} \to \infty$. We can thus apply Theorem 1 and obtain

$$\psi(G_n(\tau_n)) - \alpha_n \xrightarrow{d} \sum_{H\in\widetilde{\mathcal{U}}}\tilde{a}(H)U(H; 1). \qquad \square$$

REMARK 6.2. We will see in the next section (Remark 7.7 and Theorem 11) that α_n may be replaced by $\text{E} X_n$ in (6.9) and that, under an extra condition, actually all moments converge in (6.9). We conjecture that no extra condition is necessary to guarantee convergence of the variance, and in particular that in the case of asymptotic normality, $(X_n - \text{E} X_n)/(\text{Var} X_n)^{1/2} \xrightarrow{d} N(0, 1)$, but we have not been able to prove this.

REMARK 6.3. We say that $\psi(G_{n,m_n})$ is dominated by a family $\mathcal{H} \subseteq \widetilde{\mathcal{U}}$ if the assumptions of Theorem 7 hold with $\tilde{a}(H) = 0$ for $H \in \widetilde{\mathcal{U}} \setminus \mathcal{H}$, but not for all $H \in \mathcal{H}$. This notion will be used somewhat informally, and extended to situations as in the theorems below without further comment.

Roughly speaking, Theorem 7 says that we obtain the limit for G_{n,m_n} by ignoring all terms in the decomposition for graphs H containing a component K_2, and then taking the limit of the remainder for G_{n,p_n}. This is not completely true, however, for two reasons. First, the terms for graphs with an even number of K_2 components may give contributions in (6.10). Secondly, (6.8) imposes a bound on the coefficients also for graphs with an odd number of K_2 (except K_2 itself).

There are also cases when Theorem 7 does not give a non-degenerate limit, for example if $\psi_n(G_n(p_n))$ is dominated by $K_2 + K_2$ or by $K_2 + H$, $H \in \widetilde{\mathcal{U}}$. Such cases may be handled by modifying ψ by a suitable functional that vanishes for G_{n,m_n}, for example $(e(G) - m_n)\varphi(G)$ where φ is any graph statistic (cf. Example 6.1). We give the details for the simplest (and most important) case; another case is used in the proof of Theorem 23.

THEOREM 8. *Suppose that all conditions of Theorem 7 hold, except that* (6.8) *is not required for* $H = K_2 + K_2$. *Suppose also that*

$$n^{-1/2}\beta_n^{-1}\hat{\psi}_n^*(K_2 + K_2; p_n) = n^{3/2}p_n\hat{\psi}_n(K_2 + K_2; p_n)/\beta_n \to a' \qquad (6.31)$$

and that $n^{3/2}p_n \to \infty$. *Then*

$$\frac{X_n - \alpha_n}{\beta_n} \xrightarrow{d} \sum_{H \in \widetilde{\mathcal{U}}} \widetilde{a}'(H)U(H; 1), \qquad (6.32)$$

where $\widetilde{\mathcal{U}}$ *and* $U(H; 1)$ *are as in Theorem 7,*

$$\alpha_n = \hat{\psi}_n(\emptyset; p_n) - 2(1 - p_n)\hat{\psi}_n^*(K_2 + K_2; p_n) + \beta_n \sum_{j \geq 2}(-1)^j \frac{(2j)!}{j!}(1 - p)^j a(2jK_2)$$

$$(6.33)$$

and

$$\widetilde{a}'(H) = \begin{cases} \widetilde{a}(P_2) - 4a', & H = P_2, \\ \widetilde{a}(H), & H \neq P_2, \end{cases} \qquad (6.34)$$

where $\widetilde{a}(H)$ *is given by* (6.11).

Proof. Let $\chi(G) = e(G)(e(G) - 1)$. Then

$$\chi(G_n(t)) = \sum_{e_1 \neq e_2 \in K_n} I_{e_1}(t)I_{e_2}(t)$$

$$= \sum_{e_1 \neq e_2} I'_{e_1}(t)I'_{e_2}(t) + 2(\tbinom{n}{2} - 1)t \sum_{e \in K_n} I'_e(t) + \tbinom{n}{2}(\tbinom{n}{2} - 1)t^2$$

$$= \frac{1}{4}S_n(K_2 + K_2; t) + S_n(P_2; t) + (\tbinom{n}{2} - 1)tS_n(K_2; t) + \tbinom{n}{2}(\tbinom{n}{2} - 1)t^2.$$

Hence,

$$\hat{\chi}_n(K_2 + K_2; p_n) = \frac{1}{4}, \qquad (6.35)$$

$$\hat{\chi}_n(P_2; p_n) = 1, \qquad (6.36)$$

$$\hat{\chi}_n(\emptyset; p_n) = \tbinom{n}{2}(\tbinom{n}{2} - 1)p_n^2 = m_n(m_n - p_n). \qquad (6.37)$$

If we replace ψ by $\psi'(G) = \psi(G) + 4\hat{\psi}_n(K_2 + K_2; p_n)(m_n(m_n - 1) - \chi(G))$, where $n = v(G)$, then $\psi(G_{n,m_n}) = \psi'(G_{n,m_n})$. The result follows by applying Theorem 7 to ψ', using

$$\hat{\psi}'(K_2 + K_2; p_n) = 0, \tag{6.38}$$

$$\hat{\psi}'(P_2; p_n) = \hat{\psi}(P_2; p_n) - 4\hat{\psi}_n(K_2 + K_2; p_n), \tag{6.39}$$

$$\hat{\psi}'(\emptyset; p_n) = \hat{\psi}(\emptyset; p_n) - 4\hat{\psi}_n(K_2 + K_2; p_n)m_n(1 - p_n). \tag{6.40}$$

$\square$

We do not know whether the condition that ψ has a finite decomposition in Theorems 7 and 8 can be relaxed to e.g. finitely dominated without further restrictions. The next theorem, which can be applied to (almost) finitely dominated graph statistics (using Theorem 2 or 3), but also in other cases (using Theorem 6 or other methods for $G_{n,p}$ such as Stein's method [3]), uses a condition on the drift of $\psi(G_n(t))$.

We give two cases; the first assumes that the K_2 term in the decomposition is negligible, while the second is more general. Further cases may easily be derived, either by the method of Theorem 8, or by using a higher order Taylor expansion of α_n in the proof below. We also give pre-packed combinations with Theorems 2 and 3.

THEOREM 9. Let $m_n \to \infty$ and define $p_n = m_n / \binom{n}{2}$. Suppose that $\limsup p_n < 1$. Let ψ be a graph statistic and assume that $X_n(t) = \psi(G_n(t))$ has drift $\xi_n(t)$. Let $\alpha_n = \mathrm{E}\,\psi(G_{n,p_n})$ and let β_n be a given sequence of positive numbers.

(i) Suppose that

$$\hat{\psi}_n^*(K_2; p_n) = o(\beta_n) \tag{6.41}$$

and that for some $\gamma \geq 0$ and every sequence $t_n = p_n + O(p_n^{1/2}/n)$,

$$\mathrm{Var}(X_n(t_n)) = \mathrm{Var}(\psi(G_{n,t_n})) = (\gamma + o(1))\beta_n^2, \tag{6.42}$$

$$\mathrm{Var}(\xi_n(t_n)) = o(n^2\beta_n^2/p_n). \tag{6.43}$$

If furthermore,

$$(\psi(G_{n,p_n}) - \alpha_n)/\beta_n \xrightarrow{d} Z, \tag{6.44}$$

for some random variable Z, then

$$(\psi(G_{n,m_n}) - \alpha_n)/\beta_n \xrightarrow{d} Z. \tag{6.45}$$

(ii) Let $Y_n(t) = X_n(t) - \mathrm{E}\,X_n(t) - \hat{\psi}_n(K_2; t)S_n(K_2; t)$ and suppose that

$$\hat{\psi}_n^*(K_2 + K_2; p_n) = o(\beta_n), \tag{6.46}$$

and that, for some $\gamma \geq 0$ and every sequence $t_n = p_n + O(p_n^{1/2}/n)$,

$$\mathrm{Var}(Y_n(t_n)) = (\gamma + o(1))\beta_n^2, \tag{6.47}$$

$$\mathrm{Var}(\eta_n(t_n)) = o(n^2\beta_n^2/p_n), \tag{6.48}$$

where

$$\eta_n(t) = \xi_n(t) - \xi_n^{\{\emptyset, K_2\}}(t)$$

$$= \xi_n(t) - \mathrm{E}\,\xi_n(t) - \left(\frac{d}{dt}\hat{\psi}_n(K_2; t) - \frac{1}{1-t}\hat{\psi}_n(K_2; t)\right)S_n(K_2; t) \tag{6.49}$$

is the drift of $Y_n(t)$. If, furthermore,

$$Y_n(p_n)/\beta_n \xrightarrow{d} Z, \tag{6.50}$$

for some random variable Z, then

$$(\psi(G_{n,m_n}) - \alpha_n)/\beta_n \xrightarrow{d} Z. \tag{6.51}$$

(iii) *Suppose that the conditions of Theorem 2 are satisfied with $p < 1$, and that for every sequence $t_n = p_n + O(p_n^{1/2}/n)$, (6.42), (6.43) and (6.46) hold. Then*

$$(\psi(G_{n,m_n}) - \alpha_n)/\beta_n \xrightarrow{d} \sum_{H \in \mathcal{H} \setminus \{K_2\}} a(H)U(H; 1). \tag{6.52}$$

(iv) *Suppose that the conditions of Theorem 3 are satisfied, that $\hat{\psi}_n^*(K_2; p_n)/\beta_n \to a(K_2)$, and that for every sequence $t_n = p_n + O(p_n^{1/2}/n)$, (6.42) and (6.43) hold. Then*

$$(\psi(G_{n,m_n}) - \alpha_n)/\beta_n \xrightarrow{d} N(0, \gamma - 2a(K_2)^2(1-p)). \tag{6.53}$$

Proof. We may assume that $\beta_n = 1$, since we otherwise may divide ψ by β_n. Let $\delta_n = p_n^{1/2}/n$. We assume whenever necessary that n is large enough.

(i) Fix $\lambda > 0$ and note that (6.43) holds uniformly for $|t_n - p_n| \leq \lambda \delta_n$ since we may choose t_n in that interval such that $\mathrm{Var}(\xi_n(t_n)) \geq \frac{1}{2}\sup\{\mathrm{Var}(\xi_n(t)) : |t - p_n| \leq \lambda \delta_n\}$. In particular,

$$\int_{p_n - \lambda\delta_n}^{p_n + \lambda\delta_n} (\mathrm{Var}\,\xi_n(t))^{1/2}dt = o(\lambda\delta_n n/p_n^{1/2}) = o(1). \tag{6.54}$$

We apply Lemma 2.2(ii) to $\overline{X}_n(t) = X_n(t) - \mathrm{E}\,X_n(t) = X_n(t) - \hat{\psi}_n(\emptyset; t)$ and obtain, since $\overline{X}_n(t)$ has drift $\xi_n(t) - \mathrm{E}\,\xi_n(t)$ and (6.42) yields $\mathrm{E}(\overline{X}_n(p_n + \lambda\delta_n))^2 - \mathrm{E}(\overline{X}_n(p_n - \lambda\delta_n))^2 \to 0$,

$$\left\|\sup_{|t-p_n|\leq\lambda\delta_n} |\overline{X}(t) - \overline{X}(p_n - \lambda\delta_n)|\right\|_2 \to 0. \tag{6.55}$$

Consequently,

$$\|(\overline{X}(\tau_n) - \overline{X}_n(p_n))I(|\tau_n - p_n| \leq \lambda p_n^{1/2}/n)\|_2$$

$$\leq 2\left\|\sup_{|t-p_n|\leq\lambda\delta_n} |\overline{X}(t) - \overline{X}(p_n - \lambda\delta_n)|\right\|_2 \to 0,$$

for every $\lambda > 0$, and Lemma 6.2 yields

$$\overline{X}_n(\tau_n) - \overline{X}_n(p_n) \xrightarrow{p} 0. \tag{6.56}$$

Since (6.44) may be written $\overline{X}_n(p_n) \xrightarrow{d} Z$, this yields, with $\alpha_n(t) = \mathrm{E}\, X_n(t) = \hat{\psi}_n(\emptyset; t)$,

$$X_n(\tau_n) - \alpha_n(\tau_n) = \overline{X}_n(\tau_n) \xrightarrow{d} Z. \tag{6.57}$$

Furthermore, by Propositions 4.9 and 1.1, we have, uniformly for $|t - p_n| \le \lambda\delta_n$,

$$\left| \frac{d}{dt}\left((1-t)\hat{\psi}_n(K_2; t) \right) \right|^2 \le \mathrm{Var}(\xi_n(t))/\,\mathrm{E}\,|\check{S}_n(K_2; t)|^2 = o(p_n^{-2}), \tag{6.58}$$

and thus, using (6.41),

$$\hat{\psi}_n(K_2; t) = \tfrac{1-p_n}{1-t}\hat{\psi}(K_2; p_n) + o(p_n^{-1}\delta_n) = o(n^{-1}p_n^{-1/2}). \tag{6.59}$$

By the mean value theorem and Proposition 4.2, for some ρ between τ_n and p_n,

$$\alpha_n(\tau_n) - \alpha_n(p_n) = (\tau_n - p_n)\alpha_n'(\rho) = (\tau_n - p_n)n(n-1)\hat{\psi}(K_2; \rho). \tag{6.60}$$

Hence, if $|\tau_n - p_n| \le \lambda\delta_n$, (6.59) yields $|\alpha_n(\tau_n) - \alpha_n(p_n)| \to 0$. Lemma 6.2 yields $\alpha_n(\tau_n) - \alpha_n(p_n) \xrightarrow{p} 0$, which together with (6.57) yields $X_n(\tau_n) - \alpha_n(p_n) \xrightarrow{d} Z$.

(ii) We obtain as above

$$Y_n(\tau_n) - Y_n(p_n) \xrightarrow{p} 0 \tag{6.61}$$

and thus

$$Y_n(\tau_n) \xrightarrow{d} Z. \tag{6.62}$$

Furthermore, with $\alpha_n(t)$ as above, by Proposition 4.2 and Lemma 6.1,

$$\begin{aligned}
Y_n(\tau_n) &= X_n(\tau_n) - \alpha_n(\tau_n) - \alpha_n'(\tau_n)(n(n-1))^{-1}S_n(K_2; \tau_n) \\
&= X_n(\tau_n) - \alpha_n(\tau_n) - \alpha_n'(\tau_n)(p_n - \tau_n),
\end{aligned} \tag{6.63}$$

and thus

$$\begin{aligned}
Y_n(\tau_n) - (X_n(\tau_n) - \alpha_n(p_n)) &= \alpha_n(p_n) - \alpha_n(\tau_n) - \alpha_n'(\tau_n)(p_n - \tau_n) \\
&= \tfrac{1}{2}(p_n - \tau_n)^2\alpha_n''(\rho)
\end{aligned} \tag{6.64}$$

for some ρ between τ_n and p_n.

Let $t_n = p_n + O(p_n^{1/2}/n)$. Then, by (4.6) and (6.47), for every $H \ne \emptyset, K_2$,

$$|\hat{\psi}_n^*(H; t_n)|^2 = O(\mathrm{Var}\,\psi_n^{\{H\}}(t_n)) = O(\mathrm{Var}\, Y_n(t)) = O(1). \tag{6.65}$$

Moreover, (6.46) and (6.48) imply, by the same argument as for (6.59),

$$\hat{\psi}_n(K_2 + K_2; t_n) = o(n^{-2}p_n^{-1}). \tag{6.66}$$

Proposition 4.2, (6.65) for $H = P_2$ and (6.66) yield

$$\begin{aligned}
|\alpha''(t_n)| &= |2(n)_3 \hat{\psi}_n(P_2; t_n) + 2(n)_4 \hat{\psi}_n(K_2 + K_2; t_n)| \\
&\leq 2n^{3/2} t_n^{-1} |\hat{\psi}_n^*(P_2; t_n)| + o(n^2 p_n^{-1}) \\
&= o(n^2 p_n^{-1}).
\end{aligned} \tag{6.67}$$

Consequently, if $|\tau_n - p_n| \leq \lambda \delta_n$ and ρ is as above, $(p_n - \tau_n)^2 \alpha_n''(\rho) \to 0$. By (6.64) and Lemma 6.2 again,

$$Y_n(\tau_n) - (X_n(\tau_n) - \alpha_n(p_n)) \xrightarrow{p} 0, \tag{6.68}$$

and (6.62) yields $X_n(\tau_n) - \alpha_n(p_n) \xrightarrow{d} Z$, i.e. (6.51).

(iii) Let $t_n = p_n + O(p_n^{1/2}/n)$. Since $\mathrm{Var}(\eta_n(t_n)) \leq \mathrm{Var}(\xi_n(t_n))$, (6.43) implies (6.48). Moreover, (6.58) follows as above, and thus, using the assumption (5.2) (which holds, with $a(H) = 0$, also for $H \notin \mathcal{H}$),

$$\hat{\psi}_n^*(K_2; t_n) = np_n^{1/2} \hat{\psi}_n(K_2; p_n)(1 + o(1)) + o(1) = a(K_2) + o(1). \tag{6.69}$$

Consequently, by (6.42) and (4.6),

$$\begin{aligned}
\mathrm{Var}(Y_n(t_n)) &= \mathrm{Var}(X_n(t_n)) - (\hat{\psi}_n(K_2; t_n))^2 \mathrm{Var}(S_n(K_2; t_n)) \\
&= \gamma - 2a(K_2)^2(1 - p) + o(1), \tag{6.70}
\end{aligned}$$

and thus (6.47) holds (with a new γ). Finally, Theorem 2 applied to the modification $\psi(G) - 2\hat{\psi}_n(K_2; p_n)(e(G) - \binom{n}{2}p_n)$ yields $Y_n(p_n) \xrightarrow{d} \sum_{H \in \mathcal{H} \setminus \{K_2\}} a(H)U(H; 1)$ (the case when $a(H) = 0$ for $H \neq K_2$ is simple), and (6.52) follows by (ii).

(iv) Since there exists a dominating family not containing $K_2 + K_2$, (4.6), (4.18) and (6.42) imply $\hat{\psi}_n^*(K_2 + K_2; p_n) = o(1)$ so (6.46) holds, cf. Remarks 4.3 and 4.4. The result now follows from (iii) by taking subsequences as in the proof of Theorem 3, using e.g. (5.4) and (6.70) to calculate the variance of the limit.   □

REMARK 6.4. Joint convergence follows for any family $\psi^{(i)}$ of statistics that satisfy the conditions of one of these theorems, with common m_n but possibly different β_n; for Theorem 9 we also require joint convergence in (6.44) and (6.50). Note also that the proof of Theorem 9 shows that $(X_n(\tau_n) - \alpha_n)/\beta_n \approx \overline{X}_n(p_n)/\beta_n$ in case (i) and $\approx Y_n(p_n)/\beta_n$ in cases (ii)–(iv).

REMARK 6.5. We derive limit results for $G_{n,m}$ from results for $G_{n,p}$. In the opposite direction, Pittel [22] has proved a related result, where results for $G_{n,p}$ follow from limit results for $G_{n,m}$, again possibly with a different asymptotic variance.

7. Moment convergence

We have proved theorems on convergence in distribution of graph statistics. In this section we show that some extra conditions imply moment convergence. We consider only the case of a finite decomposition.

We say that $Z_n \overset{d}{\to} Z$ with convergence of all moments, if $Z_n \overset{d}{\to} Z$ and $E Z_n^k \to E Z^k < \infty$ for every integer $k \geq 1$; it then follows that $E |Z_n|^r \to E |Z|^r$ for every $r > 0$.

We define

$$\gamma(H, n, t) = \min_{F \subseteq H} n^{v(F)} t^{e(F)}. \tag{7.1}$$

Note that $\gamma(H, n, t) \leq n^{v(\emptyset)} t^{e(\emptyset)} = 1$ and that $\gamma(H, n, t) = 1 \iff n t^{m(H)} \geq 1$, cf. the related Remark 3.1.

LEMMA 7.1. *Let H be a graph without isolated vertices. For every $r > 0$, there exist $C = C(H, r) < \infty$, such that for all $n \geq 1$ and $t > 0$,*

$$\left\| \sup_{0 \leq s \leq t} |S_n(H; s)| \right\|_r \leq \begin{cases} C(H, r) n^{v(H)/2} t^{e(H)/2}, & r \leq 2, \\ C(H, r) n^{v(H)/2} t^{e(H)/2} \gamma(H, n, t)^{-(1/2 - 1/r)}, & r \geq 2. \end{cases} \tag{7.2}$$

Proof. It suffices to prove (7.1) when r is an even integer. The result for other values of $r > 2$ follow by interpolation ($\log \|X\|_r$ is a convex function of $1/r$ for every X) while the case $r < 2$ follows by Lyapunov's inequality ($\|X\|_r$ is increasing in r). Furthermore, as in the proof of Lemma 3.2, the result for $t \geq 1/2$ follows from the case $t = 1/2$ by time reversal, while if $t \leq 1/2$ we may use Doob's inequality for the martingale $\check{S}_n(H)$. Hence it suffices to prove that if $k \geq 1$ is an integer, then

$$E |S_n(H; t)|^{2k} \leq C \left(n^{v(H)/2} t^{e(H)/2} \gamma(H, n, t)^{-\left(\frac{1}{2} - \frac{1}{2k}\right)} \right)^{2k}. \tag{7.3}$$

In order to show this, we use the expansion

$$E |S_n(H; t)|^{2k} = \sum_{H_1, \dots, H_{2k}} E \prod_{i=1}^{2k} \prod_{e \in H_i} I_e'(t),$$

where we sum over all $H_1, \dots, H_{2k} \subseteq K_n$ that are isomorphic to H, counted with multiplicities. Consider one term in this sum and let $\tilde{H} = \bigcup_{i=1}^{2k} H_i$; let further v_j and e_j be the numbers of vertices and edges, respectively, in $\tilde{H}$ that appear in exactly j of $H_1, \dots, H_{2k}$, $j = 1, \dots, 2k$. Since the term is $O(t^{e(\tilde{H})})$, and vanishes unless $e_1 = 0$ (and thus $v_1 = 0$), it follows, arguing as in the proof of Lemma 3.1, that

$$E |S_n(H; t)|^{2k} \leq C(H, k) \max n^{v(\tilde{H})} t^{e(\tilde{H})}, \tag{7.4}$$

where the maximum is taken over all such $\tilde{H}$ with $e_1 = v_1 = 0$. Let $F_i = \bigcup_{i < l < m} H_i \cap H_l \cap H_m$. Then $\sum_1^{2k-2} v(F_i) = \sum_3^{2k} (j-2) v_j$ and $\sum_1^{2k-2} e(F_i) = \sum_3^{2k} (j-2) e_j$. Hence, if $v_1 = e_1 = 0$, $\sum_1^{2k-2} v(F_i) = \sum_1^{2k} (j-2) v_j = \sum_1^{2k} j v_j -$

$2\sum_1^{2k} v_j = \sum_1^{2k} v(H_i) - 2v(\tilde{H}) = 2kv(H) - 2v(\tilde{H})$, and similarly $\sum_1^{2k-2} e(F_i) = 2ke(H) - 2e(\tilde{H})$. This yields, since $F_i \subseteq H_i \cong H$,

$$n^{v(\tilde{H})} t^{e(\tilde{H})} = n^{kv(H)} t^{ke(H)} \prod_{i=1}^{2k-2} (n^{v(F_i)} t^{e(F_i)})^{-1/2}$$

$$\leq n^{kv(H)} t^{ke(H)} \gamma(H, n, t)^{-(2k-2)/2}, \tag{7.5}$$

which together with (7.4) yields (7.3). $\square$

REMARK 7.1. As a corollary we have for all $r > 0$ the weaker estimate

$$\| \sup_{0 \leq s \leq t} |S_n(H; s)| \|_r \leq C(H, r) n^{v(H)/2} t^{e(H)/2} \gamma(H, n, t)^{-1/2}. \tag{7.6}$$

REMARK 7.2. In fact, if $t \leq t_0 < 1$, then $\|S_n(H; t)\|_r$ is of the same order as the right hand side in (7.2) so we may replace $\leq$ by $\asymp$. For $r \leq 2$ this follows from Fatou's lemma and Theorem 1; for r an even integer it follows from the proof above and for other $r > 2$ we then use interpolation. It follows that condition (7.9) in the theorem below cannot be improved.

THEOREM 10. Let $X_n = \psi(G_{n,p_n})$, where ψ has a finite decomposition and $p_n \to p \geq 0$. Suppose that β_n is a sequence of positive numbers such that

$$\lim_{n \to \infty} \hat{\psi}_n^*(H; p_n)/\beta_n = a(H) \tag{7.7}$$

exists for every $H \neq \emptyset$, and that for every $H \neq \emptyset$, either

$$np_n^{m(H)} \to \infty \tag{7.8}$$

or, for every $\varepsilon > 0$,

$$\hat{\psi}_n^*(H; p_n) = o(\beta_n \gamma(H, n, p_n)^{1/2-\varepsilon}). \tag{7.9}$$

Then

$$\frac{X_n - \mathrm{E}\, X_n}{\beta_n} \xrightarrow{d} \sum_{H \neq \emptyset} a(H) U(H; 1) \tag{7.10}$$

with convergence of all moments, where $U(H; 1)$ is as in Theorem 1. In particular, if $a(H) = 0$ for every disconnected H, then $(X_n - \mathrm{E}\, X_n)/\beta_n \xrightarrow{d} N(0, \sigma^2)$ with convergence of all moments, where $\sigma^2 = \lim_{n \to \infty} \mathrm{Var}(X_n)/\beta_n^2$.

Proof. Note that (7.9) implies $\hat{\psi}_n^*(H; p_n) = o(\beta_n)$ and thus $a(H) = 0$. The convergence in distribution (7.10) thus follows by Theorem 1 and (4.6), or by Theorem 2 with $\mathcal{H} = \{H \neq \emptyset : a(H) \neq 0\}$ (unless $p = 1$ or every $a(H) = 0$, when (7.10) follows easily with the right hand side equal to 0).

If (7.8) holds, then $\gamma(H, n, p_n) = 1$ for large n, and thus (7.7) yields

$$\hat{\psi}_n^*(H; p_n) = O(\beta_n \gamma(H, n, p_n)^{1/2-\varepsilon}); \tag{7.11}$$

otherwise this follows directly from (7.9). Hence, by Minkowski's inequality and Lemma 7.1, for any $r \geq 2$,

$$\|(X_n - \mathrm{E}\, X_n)/\beta_n\|_r < \sum_{H \neq \emptyset} \|\hat{\psi}_n(H; p_n) S_n(H; p_n)\|_r / \beta_n$$

$$\leq \sum_{H \neq \emptyset} C(H, r) |\hat{\psi}_n^*(H; p_n)| \beta_n^{-1} \gamma(H, n, p_n)^{-(1/2 - 1/r)},$$

where the terms stay bounded as $n \to \infty$. Since it suffices to sum over a finite set of H, $\|(X_n - \mathrm{E}\, X_n)/\beta_n\|_r$ stays bounded as $n \to \infty$, and hence all moments of lower order converge. $\qquad\square$

REMARK 7.3. If $p > 0$, then (7.8) holds for every H (while (7.9) is equivalent to $a(H) = 0$). Hence, when $p > 0$ and ψ has a finite decomposition, moment convergence always holds in Theorem 2. The same is true for Theorem 3, cf. Remark 7.8 below.

REMARK 7.4. Condition (7.9) is implied by the two conditions $a(H) = 0$ and

$$\hat{\psi}_n^*(H; p_n) = O\big(\beta_n \gamma(H, n, p_n)^{1/2}\big). \tag{7.12}$$

REMARK 7.5. If (7.9) holds just for some $\varepsilon > 0$ (and every H that does not satisfy (7.8)), then all moments of order $\leq 1/\varepsilon$ converge.

There are similar results for G_{n, m_n}. We let τ_n and $p_n = m_n/\binom{n}{2}$ be as earlier.

LEMMA 7.2. If $x \geq 0$, then

$$\mathrm{P}(\tau_n > x p_n) \leq e^{1 - x/2}. \tag{7.13}$$

Proof. The result is trivial if $m_n = 0$, $x p_n \geq 1$ or $x \leq 2$. Otherwise we let $N = \binom{n}{2}$ and obtain for every $u > 0$,

$$\begin{aligned}
\mathrm{P}(\tau_n > x p_n) &= \mathrm{P}\big(e(G_n(x p_n)) < m_n\big) \\
&\leq e^{u m_n} \,\mathrm{E} \exp\big(-u e(G_n(x p_n))\big) \\
&= e^{u m_n} \big(1 - (1 - e^{-u}) x p_n\big)^N \\
&\leq e^{u m_n} \exp\big(-N(1 - e^{-u}) x p_n\big) \\
&= \exp\big(m_n(u - x(1 - e^{-u}))\big).
\end{aligned} \tag{7.14}$$

Now choose $u = 1$ to obtain $\mathrm{P}(\tau_n > x p_n) \leq \exp(m_n(1 - x/2)) \leq \exp(1 - x/2)$. $\qquad\square$

LEMMA 7.3. *Let H be a graph without isolated vertices. For every $r > 0$ and $\varepsilon > 0$, there exist constants $C(H, r)$, $C(H, r, \varepsilon) < \infty$ such that*

$$\|S_n(H; \tau_n)\|_r \leq \begin{cases} C(H, r) n^{v(H)/2} p_n^{e(H)/2}, & r < 2, \\ C(H, r, \varepsilon) n^{v(H)/2} p_n^{e(H)/2} \gamma(H, n, p_n)^{-(1/2 - 1/r + \varepsilon)}, & r \geq 2. \end{cases}$$

Proof. The case $m_n = 0$ is trivial. Otherwise, let $M(t) = \sup_{s \leq t} |S_n(H; s)|$. Then

$$S_n(H; \tau_n) \leq \sum_{j=1}^{\infty} M(jp_n)I((j-1)p_n < \tau_n \leq jp_n)$$

and hence, if $r \geq 1$ (which we may assume) and $1/r = 1/r_1 + 1/q$ with $2 \leq r_1 < \infty$ and $q < \infty$,

$$\|S_n(H; \tau_n)\|_r \leq \sum_{j=1}^{\infty} \|M(jp_n)\|_{r_1} \|I(\tau_n > (j-1)p_n)\|_q$$

$$\leq C_1 \sum_{j=1}^{\infty} n^{v(H)/2}(p_n j)^{e(H)/2} \gamma(H, n, p_n j)^{-(1/2-1/r_1)} e^{-(j-3)/2q}$$

$$\leq C_2 n^{v(H)/2} p_n^{e(H)/2} \gamma(H, n, p_n)^{-(1/2-1/r_1)}. \tag{7.15}$$

by Minkowski's and Hölder's inequalities and Lemma 7.1. The result follows by choosing $r_1 = 2$ when $r < 2$ and $1/r_1 = 1/r - \varepsilon$ (assuming $\varepsilon < 1/r$) when $r \geq 2$. $\qquad\square$

THEOREM 11. *Let* $X_n = \psi(G_{n,m_n})$ *where* ψ *has a finite decomposition,* $m_n \to \infty$ *and* $p_n = m_n/\binom{n}{2} \to p < 1$. *Suppose that* β_n *is a sequence of positive numbers such that*

$$\hat{\psi}_n^*(H; p_n)/\beta_n \to a(H) \tag{7.16}$$

for every $H \in \mathcal{U}^0$ *with at least three vertices and some numbers* $a(H)$, *and that, for every such graph* H, *either*

$$np_n^{m(H)} \to \infty \tag{7.17}$$

or, for every $\varepsilon > 0$,

$$\hat{\psi}_n^*(H; p_n) = o(\beta_n \gamma(H, n, p_n)^{1/2-\varepsilon}). \tag{7.18}$$

Then, with notation as in Theorem 7,

$$\frac{X_n - \mathrm{E}\,X_n}{\beta_n} \xrightarrow{d} \sum_{H \in \widetilde{\mathcal{U}}} \tilde{a}(H)U(H; 1) \tag{7.19}$$

with convergence of all moments. Furthermore, $\mathrm{E}\,X_n = \alpha_n + o(\beta_n)$ *and*

$$\mathrm{Var}\,X_n/\beta_n^2 \to \sum_{H \in \widetilde{\mathcal{U}}} \tilde{a}(H)^2 \operatorname{aut}(H)(1-p)^{e(H)}. \tag{7.20}$$

In particular, if $a(H) = 0$ *for every* H *that has two or more components with three or more vertices, and* $\tilde{a}(H) \neq 0$ *for some* $H \in \widetilde{\mathcal{U}}$, *then*

$$(X_n - \mathrm{E}\,X_n)/(\mathrm{Var}\,X_n)^{1/2} \xrightarrow{d} N(0, 1) \tag{7.21}$$

with convergence of all moments.

Proof. We may assume that $\beta_n = 1$. We may also assume, by first modifying ψ as in Example 6.1, subtracting $2\hat{\psi}_n(K_2; p_n)(e(G) - m_n)$ from $\psi(G)$ (with $n = v(G)$), that $\hat{\psi}_n(K_2; p_n) = 0$, and further, by subtracting the constant $\hat{\psi}_n(\emptyset, p_n)$, that $\hat{\psi}_n(\emptyset, p_n) = 0$. Then, (7.16) and (7.17) or (7.18) hold for every $H \in \mathcal{U}^0$, which as in the proof of Theorem 10 yields (7.11), i.e.

$$\hat{\psi}_n^*(H; p_n) = O(n^{-v(H)/2} p_n^{-e(H)/2} \gamma(H, n, p_n)^{1/2-\varepsilon}). \qquad (7.22)$$

If $H' \supseteq H$, then $\gamma(H', n, p_n) \leq \gamma(H, n, p_n)$, and thus Proposition 4.3 and induction, see (6.13), yield by (7.22)

$$\frac{d^k}{dt^k} \hat{\psi}_n^*(H; p_n) = O(n^{-v(H)/2+k} p_n^{-e(H)/2-k/2} \gamma(H, n, p_n)^{1/2-\varepsilon}) \qquad (7.23)$$

for all $H \in \mathcal{U}^0$ and $k \geq 0$ (assuming $0 < \varepsilon < 1/2$). Proposition 4.5 and Lemma 6.1 then yield

$$|\hat{\psi}_n(H; \tau_n)| = |\sum_{k=0}^{d} \frac{1}{k!} \frac{d^k}{dt^k} \hat{\psi}_n(H; p_n)(\tau_n - p_n)^k|$$

$$\leq C_1 \sum_{k=0}^{d} n^{-v(H)/2+k} p_n^{-e(H)/2-k/2} |\tau_n - p_n|^k \gamma(H, n, p_n)^{1/2-\varepsilon}$$

$$\leq C_2 \sum_{k=0}^{d} n^{-v(H)/2-k} p_n^{-e(H)/2-k/2} |S_n(K_2; \tau_n)|^k \gamma(H, n, p_n)^{1/2-\varepsilon}.$$

Hence, for every $r \geq 1$ (with C_j depending on H, ε and possibly r),

$$\|\hat{\psi}_n(H; \tau_n)\|_r \leq C_2 \sum_{k=0}^{d} n^{-v(H)/2-k} p_n^{-e(H)/2-k/2} \|S_n(K_2; \tau_n)\|_{kr}^k \gamma(H, n, p_n)^{1/2-\varepsilon}$$

$$\leq C_3 n^{-v(H)/2} p_n^{-e(H)/2} \gamma(H, n, p_n)^{1/2-\varepsilon} \qquad (7.24)$$

by Lemma 7.3, because $\gamma(K_2, n, p_n) = 1$ whenever $m_n \geq 1$. The decomposition (6.6) and Minkowski's and Hölder's inequalities now yield, for every $r \geq 1$, using (7.24) and Lemma 7.3 again,

$$\|X_n - \alpha_n\|_r \leq \sum_{H \in \mathcal{U}^0} \|\hat{\psi}_n(H; \tau_n)\|_{2r} \|S_n(H; \tau_n)\|_{2r} + |\alpha_n| \leq C. \qquad (7.25)$$

Hence all moments are bounded, and since Theorem 7 yields the convergence in distribution, it follows that $X_n - \alpha_n \xrightarrow{d} Z$ with convergence of all moments, where $Z = \sum \tilde{a}(H) U(H; 1)$. In particular, $\mathrm{E}\, X_n - \alpha_n = \mathrm{E}(X_n - \alpha_n) \to \mathrm{E}\, Z = 0$ and $\mathrm{Var}\, X_n \to \mathrm{Var}\, Z$, and the conclusions follow. $\qquad \square$

REMARK 7.6. Similarly, in the situation of Theorem 8, (7.18) for every $H \in \mathcal{U}^0$ with at least three vertices that does not satisfy (7.17), implies moment convergence. (Use the proof of Theorem 8.)

REMARK 7.7. If (7.18) holds for some $\varepsilon \leq 1/2$ only (and every H that does not satisfy (7.17)) then all moments of order less than $1/\varepsilon$ converge. (The same proof, using other exponents in Hölder's inequality in (7.25).) In particular, the conditions of Theorem 7 ($\varepsilon = 1/2$) are sufficient to imply $\mathrm{E}\,X_n = \alpha_n + o(\beta_n)$, so α_n may be replaced by $\mathrm{E}\,X_n$ in that theorem. (Similarly, α_n may be replaced by $\mathrm{E}\,X_n$ in Theorem 8.) We do not, however, know whether the conditions of Theorem 7 imply that the variance converges (although we conjecture so). By the above, it is sufficient that also (7.18) holds for some $\varepsilon < 1/2$, i.e. $\hat{\psi}_n^*(H; p_n) = o(\beta_n \gamma(H, n, p_n)^\delta)$ for some $\delta > 0$ and every H that does not satisfy (7.17).

REMARK 7.8. As usual, the conditions can be simplified in the normal case. If $p < 1$ and $\hat{\psi}_n^*(H; p_n)/\beta_n \to 0$ for every disconnected $H \in \mathcal{U}^0$, then (7.7) may be replaced by $\mathrm{Var}(X_n)/\beta_n^2 \to \sigma^2$, and similarly for Theorem 11.

8. Functional convergence

In the last sections we have studied convergence of graph statistics $\psi(G_n(t))$ at a simple (possibly random) epoch t. Results on the joint distribution of at two or several epochs may be obtained by the same methods. In this section, we go one step further and consider the whole evolution. We will give results on convergence (in the Skorokhod topology) of the process $\psi(G_n(t))$, properly normalized. Convergence in distribution then follows for any functional of the whole process that is continuous in the Skorokhod topology, for example the maximum, which will be treated in detail in the next section.

We assume, as before, that p_n is a given sequence of numbers in $(0,1)$, and that ψ is a graph statistic. We define, for convenience,

$$a_n(H; t) = n^{v(H)/2} p_n^{e(H)/2} \hat{\psi}_n(H; p_n t). \tag{8.1}$$

Thus $a_n(H; 1) = \hat{\psi}_n^*(H; p_n)$.

We begin with the case of finite decompositions.

THEOREM 12. Let $X_n(t) = \psi(G_n(t))$, where ψ has a finite decomposition with support $\mathcal{H}$. Suppose that $p_n \to p \geq 0$ and that β_n is a sequence of positive numbers such that for every non-empty $H \in \mathcal{H}$,

$$n p_n^{m(H)} \to \infty \tag{8.2}$$

and, for some functions $a(H; t)$ and every fixed $t \geq 0$,

$$a_n(H; t)/\beta_n \to a(H; t). \tag{8.3}$$

Then, with $U(H; t)$ as in Theorem 1,

$$\frac{X_n(p_n t) - \mathrm{E}\,X_n(p_n t)}{\beta_n} \xrightarrow{d} \sum_{H \in \mathcal{H} \setminus \{\emptyset\}} a(H; t) U(H; t) \tag{8.4}$$

in $D([0,\infty))$. If $p > 0$, the convergence holds in $D([0,\infty])$ as well.

Proof. We may, as usual, assume that $\beta_n = 1$. By Proposition 4.5, all $a_n(H;t)$ are polynomials of degree $\leq d$, for $t \in [0, 1/p_n]$ and some $d < \infty$; hence the convergence (8.3) for any fixed $d+1$ values of $t \in (0, 1/p)$ implies that the coefficients of the polynomials converge. Consequently, each $a(H;t)$ is a polynomial on $[0, 1/p)$ and $a_n(H;t) \to a(H;t)$ uniformly on compact subsets. If $p > 0$, it follows easily that $a_n(H;t) \to a(H;t)$ uniformly on $[0,\infty]$, with $a(H;t)$ constant on $[1/p, \infty]$. Theorem 1 and Proposition 2.2 now yield

$$X_n(t_n) - \mathrm{E}\, X_n(t) = \sum_{\mathcal{H}\setminus\{\emptyset\}} a_n(H;t) n^{-v(H)/2} p_n^{-e(H)/2} S_n(H; p_n t)$$

$$\xrightarrow{d} \sum_{\mathcal{H}\setminus\{\emptyset\}} a(H;t) U(H;t). \qquad \square$$

As usual, we can relax the conditions somewhat when the limit is normal.

THEOREM 13. *Let $X_n(t) = \psi(G_n(t))$, where ψ has a finite decomposition with support $\mathcal{H}$. Suppose that (8.2) holds for every non-empty $H \in \mathcal{H}$, and that β_n is a sequence of positive numbers such that*

$$\varphi(s,t) = \lim_{n\to\infty} \mathrm{Cov}(X_n(p_n s), X_n(p_n t))/\beta_n^2 \qquad (8.5)$$

exists for every $s, t \geq 0$, and

$$a_n(H;t)/\beta_n \to 0 \qquad (8.6)$$

for every disconnected $H \in \mathcal{H}$ and $t > 0$. Then

$$\frac{X_n(p_n t) - \mathrm{E}\, X_n(p_n t)}{\beta_n} \xrightarrow{d} Z(t) \qquad (8.7)$$

in $D([0,\infty))$, where $Z(t)$ is a continuous Gaussian process with $\mathrm{E}\, Z(t) = 0$ and covariance function

$$\mathrm{E}(Z(s)Z(t)) = \varphi(s,t).$$

If $p > 0$ the convergence holds in $D([0,\infty])$ as well.

Proof. Assume $\beta_n = 1$. The assumption (8.5) implies that $\mathrm{Var}\, X_n(p_n t) = O(1)$ for every fixed t, and hence, by (4.2), $a_n(H;t) = O(1)$ for every t with $0 < t < 1/p$. Let d be as in the proof of Theorem 12, and choose $d+1$ numbers $t_1, \ldots, t_{d+1}$ in $(0, 1/p)$. We may for every subsequence $\{n_k\}$ select a subsubsequence for which $a_n(H;t_i)$ converges for every $H \in \mathcal{H}$ and every t_i. As in the proof of Theorem 12, this implies that, along the subsubsequence, (8.3) holds uniformly, and thus (8.4) holds. The right hand side of (8.4) is a continuous Gaussian process $Z(t)$, and using (3.3), (8.1) and (1.11), for $0 \leq s \leq t < \infty$,

$$\mathrm{Cov}(Z(s), Z(t)) = \sum_{H \neq \emptyset} a(H;s) a(H;t) \operatorname{aut}(H)(s(1-pt)_+)^{e(H)}$$

$$= \lim_{n\to\infty} \sum \hat{\psi}_n(H; p_n s)\hat{\psi}_n(H; p_n t)(n)_{v(H)} \operatorname{aut}(H)(p_n s(1 - p_n t)_+)^{e(H)}$$

$$= \lim_{n\to\infty} \mathrm{Cov}(X_n(p_n s), X_n(p_n t)) = \varphi(s,t).$$

Thus Z has the same covariance function, and thus the same distribution, for every such subsubsequence, and the result follows.   □

We can extend the results to infinite decompositions if we add a condition on the drift. We give one such result, leaving the version corresponding to Theorem 13 to the reader.

THEOREM 14. *Let $X_n(t) = \psi(G_n(t))$ and assume that $X_n(p_n t)$ is almost finitely dominated by a family $\mathcal{H}$ of non-empty graphs for every fixed $t > 0$. Suppose that $p_n \to p \geq 0$, and that β_n is a sequence of positive numbers such that, for every $H \in \mathcal{H}$, $n p_n^{m(H)} \to \infty$ and*

$$a_n(H;t)/\beta_n \to a(H;t), \qquad t \geq 0. \tag{8.8}$$

(i) *If $\xi_n(t)$ is the drift of $X_n(t)$, and there exists a continuous function $\varphi(t)$ on $[0, 1/p)$ such that*

$$p_n^2 \operatorname{Var}(\xi_n(p_n t)) \leq \beta_n^2 \varphi(t)^2, \qquad 0 < t < 1/p, \tag{8.9}$$

(note that $p_n \xi_n(p_n t)$ is the drift of $X_n(p_n t)$), then

$$\frac{X_n(p_n t) - \operatorname{E} X_n(p_n t)}{\beta_n} \xrightarrow{d} Z(t) = \sum_{\mathcal{H}} a(H;t) U(H;t) \tag{8.10}$$

in $D([0, 1/p))$. The sum on the right hand side converges in L^2, uniformly on compact subsets of $[0, 1/p)$, and $Z(t)$ is a continuous stochastic process on $[0, 1/p)$.

(ii) *If, further, (8.9) holds for all $t > 0$ and $\int_0^\infty \varphi(t) dt < \infty$, then (8.10) holds in $D([0, \infty])$ as well. In this case, $Z(t)$ is continuous on $[0, \infty]$ and $Z(t) = 0$ when $1/p \leq t \leq \infty$.*

REMARK 8.1. In (i), it is sufficient that (8.9) holds for $t \leq t_0$ and $n \geq n(t_0)$, for each $t_0 < 1/p$; for example, for $p_n t < 1 - \varepsilon$ for every $\varepsilon > 0$.

Proof. We may again assume that $\beta_n = 1$. By Proposition 4.9, for $0 < t < 1$,

$$\xi_n(t) = \sum_H \frac{d}{dt}(\hat{\psi}_n(H;t)(1-t)^{e(H)})(1-t)^{-e(H)} S_n(H;t),$$

and thus, using Proposition 1.1,

$$\operatorname{Var}(\xi_n(t)) = \sum_H |\frac{d}{dt}(\hat{\psi}_n(H;t)(1-t)^{e(H)})|^2 (1-t)^{-2e(H)} \operatorname{Var}(S_n(H;t))$$

$$\geq \sum_{H \neq \emptyset} |\frac{d}{dt}(\hat{\psi}_n(H;t)(1-t)^{e(H)})|^2 (n)_{v(H)} t^{e(H)}. \tag{8.11}$$

Let $t_0 < 1/p$ and consider first only $t \in [0, t_0]$ and n that are so large that $p_n t_0 < 1$. Defining $b_n(H;t) = a_n(H;t)(1 - p_n t)^{e(H)}$, we obtain from (8.1) and (8.11) that, for every $H \in \mathcal{U}^0 \setminus \{\emptyset\}$ and large enough n,

$$|\frac{d}{dt} b_n(H;t)|^2 = n^{v(H)} p_n^{e(H)} |\frac{d}{dt}(\hat{\psi}_n(H;p_n t)(1 - p_n t)^{e(H)})|^2$$

$$\leq \frac{n^{v(H)}}{(n)_{v(H)}} t^{-e(H)} p_n^2 \operatorname{Var}(\xi_n(p_n t)) \leq C t^{-e(H)}, \qquad t \leq t_0. \tag{8.12}$$

Consequently, for fixed $H \neq \emptyset$ and $\delta > 0$, $\frac{d}{dt} b_n(H;t)$ is uniformly bounded on $[\delta, t_0]$. Since $b_n(H;t) \to a(H;t)(1-pt)^{e(H)}$ pointwise by (8.8), it follows that $b_n(H;t) \to a(H;t)(1-pt)^{e(H)}$ uniformly on $[\delta, t_0]$, and thus $a_n(H;t) \to a(H;t)$ uniformly on $[\delta, t_0]$; in particular, $a(H;t)$ is continuous on $[\delta, t_0]$. Since $\delta > 0$ and $t_0 < 1/p$ are arbitrary, this yields

$$a_n(H;t) \to a(H;t) \qquad \text{in } C((0, 1/p)). \tag{8.13}$$

Now define, for $N \geq 1$, $H(N) = \{H \in \mathcal{H} : v(H) \leq N\}$ and $X_n^{(N)}(t) = \psi^{\mathcal{H}(N)}(G_n(t))$. We claim that, if $t_0 < 1/p$ as above,

$$X_n^{(N)}(p_n t) \xrightarrow{d} Z^{(N)}(t) = \sum_{H(N)} a(H;t) U(H;t) \qquad \text{as } n \to \infty, \tag{8.14}$$

in $D([0, 1/p))$, for every N, with $Z^{(N)}(t) \in C([0, 1/p))$ a.s.,

$$\lim_{N \to \infty} \limsup_{n \to \infty} \mathrm{E} \sup_{t \leq t_0} |X_n(p_n t) - \mathrm{E} X_n(p_n t) - X_n^{(N)}(p_n t)|^2 = 0 \tag{8.15}$$

and

$$\mathrm{E} \sup_{t \leq t_0} |Z^{(M)}(t) - Z^{(N)}(t)|^2 \to 0 \qquad \text{as } M, N \to \infty. \tag{8.16}$$

We postpone the proofs of these claims for a moment, and see how they imply the result.

Since (8.16) implies that $\{Z^{(N)}(t)\}_N$ is a Cauchy sequence in L^2 for every $t \leq t_0$, it follows that $Z^{(N)}(t) \to Z(t)$ as $N \to \infty$ in L^2, uniformly on $[0, t_0]$. Furthermore, we can select a subsequence N_i such that $Z^{(N_i)}(t)$ a.s. converges uniformly on $[0, t_0]$, and thus $Z(t)$ is a.s. continuous on $[0, t_0]$. It follows that

$$\mathrm{E} \sup_{t \leq t_0} |Z^{(N)}(t) - Z(t)|^2 \to 0 \qquad \text{as } N \to \infty, \tag{8.17}$$

and thus $Z^{(N)}(t) \to Z(t)$ in $D([0, t_0])$. Together with [4, Theorem 4.2], (8.14), (8.17) and (8.15) imply $X_n(p_n t) - \mathrm{E} X_n(p_n t) \to Z(t)$ in $D([0, t_0])$. Since t_0 is arbitrary, the convergence holds in $D([0, 1/p))$ by Proposition 2.3(ii), which is the sought result (8.10).

In order to prove (8.14), we observe that (8.13) and Theorem 1 yield

$$X_n^{(N)}(p_n t) = \sum_{\mathcal{H}(N)} a_n(H;t) n^{-v(H)/2} p_n^{-e(H)/2} S_n(H; p_n t) \xrightarrow{d} Z^{(N)}(t) \tag{8.18}$$

in $D((0, 1/p))$. Furthermore, by (8.12), for every $H \in \mathcal{H}$ and t_0, with C_j here and below not depending on n or $t \leq t_0$ (but possibly on H, t_0, N),

$$|\frac{d}{dt} b_n(H;t)| \leq C_1 t^{-e(H)/2}.$$

Hence (rather crudely), using (8.8) to infer $|b_n(H;t_0)| \leq |a_n(H;t_0)| \leq C_2$,

$$|b_n(H;t)| \leq C_3 t^{1/2-e(H)/2} + |b_n(H;t_0)| \leq C_4 t^{1/2-e(H)/2}$$

and thus

$$|a_n(H;t)| \leq C_5 t^{1/2-e(H)/2}.$$

Consequently, by Proposition 1.1,

$$\mathrm{E}\,|X_n^{(N)}(p_n t)|^2 \leq \sum_{\mathcal{H}(N)} \mathrm{aut}(H)|a_n(H;t)|^2 t^{e(H)} \leq C_6 t, \qquad t \leq t_0. \tag{8.19}$$

Since $X_n^{(N)}(p_n t)$ has drift $p_n \xi_n^{\mathcal{H}(N)}(p_n t)$ by Proposition 4.9, and

$$\mathrm{E}\,|p_n \xi_n^{\mathcal{H}(N)}(p_n t)|^2 \leq p_n^2 \,\mathrm{Var}\big(\xi_n(p_n t)\big) \leq \varphi(t)^2,$$

Lemma 2.2(i) yields

$$\mathrm{E}\sup_{s\leq t}|X_n^{(N)}(p_n s)|^2 \leq 13\,\mathrm{E}\,|X_n^{(N)}(p_n t)|^2 + 13\Big(\int_0^t \varphi(s)ds\Big)^2 \leq C_7 t. \tag{8.20}$$

By Proposition 2.4, (8.18) and (8.20) imply $X_n^{(N)} \to Z^{(N)}$ in $D([0,1/p])$.

In order to prove (8.15), we observe that, by (8.19), $\mathrm{Var}\big(X_n^{(N)}(p_n t)\big)$ is bounded for every N and t. Since $X_n(p_n t)$ is almost finitely dominated by $\mathcal{H}$, it follows from (4.21) that, for every fixed $t < 1/p$, $\mathrm{Var}\,X_n(p_n t)$ is bounded and

$$\limsup_n \mathrm{Var}\Big(X_n(p_n t) - X_n^{(N)}(p_n t)\Big) \to 0 \qquad \text{as } N \to \infty. \tag{8.21}$$

Let

$$Y_{nN}(t) = X_n(p_n t) - \mathrm{E}\,X_n(p_n t) - X_n^{(N)}(p_n t),$$

which is a semimartingale with drift $\eta_{nN}(t) = p_n(\xi_n(p_n t) - \mathrm{E}\,\xi_n(p_n t) - \xi_n^{\mathcal{H}(N)}(p_n t))$ by Proposition 4.9. Note that (8.9) implies

$$\mathrm{E}\,|\eta_{nN}(t)|^2 \leq p_n^2\,\mathrm{Var}(\xi_n(p_n t)) \leq \varphi(t)^2, \qquad t < 1/p. \tag{8.22}$$

Let $0 = t_1 < t_2 < \cdots < t_{m+1} = t_0$ be a partition of $[0,t_0]$. Then Lemma 2.2(i) yields, using (8.22),

$$\mathrm{E}\sup_{t\leq t_0}|Y_{nN}(t)|^2 \leq \sum_{i=1}^m \mathrm{E}\sup_{t_i\leq t\leq t_{i+1}}|Y_{nN}(t)|^2$$
$$\leq \sum_i 13\,\mathrm{E}\,|Y_{nN}(t_{i+1})|^2 + \sum_i 13\Big(\int_{t_i}^{t_{i+1}}\varphi(t)dt\Big)^2. \tag{8.23}$$

Taking the lim sup as first $n \to \infty$ and then $N \to \infty$, we obtain by (8.21),

$$\limsup_{N \to \infty} \limsup_{n \to \infty} \operatorname{E} \sup_{t \leq t_0} |Y_{nN}(t)|^2 \leq 13 \sum_i \left(\int_{t_i}^{t_{i+1}} \varphi(t)\, dt \right)^2$$

$$\leq 13 \int_0^{t_0} \varphi(t)\, dt \, \max_i \int_{t_i}^{t_{i+1}} \varphi(t) dt. \qquad (8.24)$$

Since the right hand side of (8.24) may be made arbitrarily small by choosing the partition $\{t_i\}$ suitably fine, the left hand side vanishes, which is (8.15).

Finally, for any fixed N and M, it follows as for (8.14) that $Y_{nN}(t) - Y_{nM}(t) = X_n^{(M)}(p_n t) - X_n^{(N)}(p_n t) \xrightarrow{d} Z^{(M)}(t) - Z^{(N)}(t)$ in $D([0, t_0])$. Since $f \to \sup |f|$ is a continuous functional on $D([0, t_0])$, this yields

$$\sup_{t \leq t_0} |Y_{nN}(t) - Y_{nM}(t)| \xrightarrow{d} \sup_{t \leq t_0} |Z^{(M)}(t) - Z^{(N)}(t)|$$

and thus, by Fatou's inequality,

$$\operatorname{E} \sup_{t \leq t_0} |Z^{(M)}(t) - Z^{(N)}(t)|^2 \leq \liminf_{n \to \infty} \operatorname{E} \sup_{t \leq t_0} |Y_{nN}(t) - Y_{nM}(t)|^2$$

$$\leq 2 \limsup_{n \to \infty} \left(\operatorname{E} \sup_{t \leq t_0} |Y_{nN}(t)|^2 + \operatorname{E} \sup_{t \leq t_0} |Y_{nM}(t)|^2 \right),$$

which together with (8.15) yields (8.16).

This completes the proof of (i).

For (ii) we first consider the case $p = 0$. For any $t_0 < \infty$ and n so large that $p_n t_0 < 1$, Lemma 2.2(i) and (8.9) yield, since $X_n(t) = \operatorname{E} X_n(t)$ for $t \geq 1$,

$$\operatorname{E} \sup_{t_0 \leq t} |X_n(p_n t) - \operatorname{E} X_n(p_n t)|^2 = \operatorname{E} \sup_{t_0 \leq t \leq 1/p_n} |X_n(p_n t) - \operatorname{E} X_n(p_n t)|^2$$

$$\leq 13 \left(\int_{t_0}^{1/p_n} \varphi(t) dt \right)^2 \leq 13 \left(\int_{t_0}^{\infty} \varphi(t) dt \right)^2.$$

The convergence in $D([0, \infty])$ now follows from Proposition 2.4.

The proof for $p > 0$ is similar, using Lemma 2.2(i) on the interval $(1/p - \delta, 1/p + \delta)$ and letting first $n \to \infty$ and then $\delta \to \infty$, finally invoking Proposition 2.3(iii) as in the proof of Theorem 1. $\qquad \square$

The version of the random graph process where edges are added at fixed times may, as stated in the introduction, be obtained from our process by a (random) change of time scale, viz. $\{G_n(T_{(m)})\}_{m=0}^{\binom{n}{2}}$. Using the methods of Section 6, we can obtain functional limit theorems for this process as well, for example the following. We let $T_{(t)} = T_{(m)}$ if $m \leq t < m + 1$ and $T_{(t)} = T_{\binom{n}{2}}$ if $t > \binom{n}{2}$.

THEOREM 15. Let $X_n(t) = \psi(G_n(T_{(\binom{n}{2})t}))$, where ψ has a finite decomposition. Suppose that $p_n \to p$ and that $np_n^{m(H)} \to \infty$ for every H in the support *of ψ. Suppose further that β_n is a sequence of positive numbers such that for every $H \in \mathcal{U}^0$ with $v(H) \geq 3$,*

$$a_n(H;t)/\beta_n \to a(H;t), \qquad t > 0, \tag{8.25}$$

for some function $a(H;t)$. Then

$$\frac{X_n(p_n t) - \alpha_n(t)}{\beta_n} \xrightarrow{d} \sum_{H \in \widetilde{\mathcal{U}}} \tilde{a}(H;t)U(H;t), \tag{8.26}$$

in $D([0,\infty))$, where $\widetilde{\mathcal{U}}$ is as in Theorem 7,

$$\alpha_n(t) = \hat{\psi}_n(0; p_n t) + \beta_n \sum_{j \geq 1} (-1)^j \frac{(2j)!}{j!}(1 - tp)_+^j a(2jK_2; t) \tag{8.27}$$

and

$$\tilde{a}(H;t) = \sum_{j \geq 0} (-1)^j \frac{(2j)!}{j!}(1 - tp)_+^j a(H + 2jK_2; t). \tag{8.28}$$

If $p > 0$, the convergence holds in $D([0,\infty])$.

Proof. This follows by an inspection of the proof of Theorem 7, checking that all error terms are uniformly small for $0 \leq t \leq t_0$, for any fixed t_0, and using the fact that if $Z_n(t) \xrightarrow{d} Z(t)$ and $\tau(t) \xrightarrow{p} t$ in $D[0,\infty)$ with Z a.s. continuous, then $Z_n(\tau(t)) \xrightarrow{d} Z(t)$ by the continuous mapping theorem (this is a functional version of Proposition 2.1). We omit the tedious details. $\square$

REMARK 8.2. Again, joint convergence follows for families of statistics satisfying the conditions of these theorems.

9. The maximum

In this section we consider the maximum $\max_{t \geq 0} \psi(G_n(t))$ of a graph statistic over the entire evolution of a random graph. Note that the time scale is irrelevant here; we may as well write this random variable as $\max_{t \geq 0} \psi(G_n(p_n t))$ or $\max_{t \geq 0} \psi(G_n(T_{(m_n t)}))$.

Since the mapping $f \to \sup_I f$ is continuous on $D(I)$ when I is a closed interval (but not otherwise), the results in the last section on convergence in $D([0,\infty])$ imply limit theorems for $\sup(X_n(t) - \mathrm{E}\,X_n(t))$, where $X_n(t) = \psi(G_n(t))$. If $\mathrm{E}\,X_n(t)$ vanishes (or is sufficiently small), this yields the sought results, see e.g. Example 12.3.

In most applications, however, $\mathrm{E}\,X_n(t)$ is much larger than $|X_n(t) - \mathrm{E}\,X_n(t)|$, and a further step is needed. The following theorem shows that in many cases the result is quite simple.

THEOREM 16. *Suppose that $X_n(t)$ are stochastic processes on $[0,\infty]$, that p_n, α_n and β_n are positive numbers and that $f(t)$ is a continuous function on $[0,\infty]$ such that*

$$\frac{X_n(p_n t) - \alpha_n f(t)}{\beta_n} \xrightarrow{d} Z(t) \qquad \text{as } n \to \infty \tag{9.1}$$

in $D([0, \infty])$, for some stochastic process $Z(t)$. Suppose further that $\alpha_n/\beta_n \to \infty$, that f has a unique maximum at t_0, and that Z is a.s. continuous at t_0. Then

$$(\sup_{t \geq 0} X_n(t) - X_n(p_n t_0))/\beta_n \xrightarrow{p} 0 \qquad (9.2)$$

and thus

$$\frac{\sup_t X_n(t) - \alpha_n f(t_0)}{\beta_n} \xrightarrow{d} Z(t_0). \qquad (9.3)$$

In particular, this yields the limit distribution of $\max \psi(G_n(t))$ whenever the conditions of either Theorem 12, 13 or 15 with $p_n = 1$, or of Theorem 14(ii) with any p_n, are satisfied, and furthermore, for some continuous f with a unique maximum, (for Theorems 12, 13, 14)

$$\sup_t |\operatorname{E} X_n(p_n t) - \alpha_n f(t)|/\beta_n \to 0,$$

or (for Theorem 15)

$$\sup_t |\alpha_n(t) - \alpha_n f(t)|/\beta_n \to 0.$$

Proof. We may assume $\beta_n = 1$ and thus $\alpha_n \to \infty$. We may also replace $X_n(t)$ by $X_n(p_n t)$ and thus assume that $p_n = 1$.

Define $E_n = \{t : f(t) > f(t_0) - \alpha_n^{-1/2}\}$. Since $[0, \infty]$ is compact, and f has a unique maximum at t_0, it is easily seen that if t_n is any point in E_n, then $t_n \to t_0$ as $n \to \infty$. Consequently, if we let $\tau_n \in E_n$ be such that

$$X_n(\tau_n) > \sup_{t \in E_n} X_n(t) - 1/n,$$

then $\tau_n \xrightarrow{p} t_0$, and Proposition 2.1 shows that

$$X_n(\tau_n) - \alpha_n f(\tau_n) - (X_n(t_0) - \alpha_n f(t_0)) \xrightarrow{p} 0.$$

Hence, for any $\varepsilon > 0$, using $f(\tau_n) \leq f(t_0)$,

$$\operatorname{P}\big(\sup_{t \in E_n} X_n(t) - X_n(t_0) > \varepsilon\big) \leq \operatorname{P}\big(X_n(\tau_n) - X_n(t_0) > \varepsilon - 1/n\big)$$

$$\leq \operatorname{P}\big(X_n(\tau_n) - \alpha_n f(\tau_n) - X_n(t_0) + \alpha_n f(t_0) > \varepsilon - 1/n\big) \to 0. \quad (9.4)$$

Turning our attention to t outside E_n, we note that if $t \notin E_n$ is such that $X_n(t) > X_n(t_0)$, then

$$\alpha_n^{1/2} \leq \alpha_n f(t_0) - \alpha_n f(t) < X_n(t) - \alpha_n f(t) - (X_n(t_0) - \alpha_n f(t_0))$$
$$\leq 2 \sup_{s \geq 0} |X_n(s) - \alpha_n f(s)|.$$

Consequently,

$$\operatorname{P}\big(\sup_{t \notin E_n} X_n(t) > X_n(t_0)\big) \leq \operatorname{P}\big(\sup_t |X_n(t) - \alpha_n f(t)| \geq \tfrac{1}{2}\alpha_n^{1/2}\big) \to 0, \qquad (9.5)$$

because $\sup_t |X_n(t) - \alpha_n f(t)| \xrightarrow{d} \sup_t |Z_t|$ and $\alpha_n^{1/2} \to \infty$.

The estimates (9.4) and (9.5), together with the trivial $\sup X_n(t) \geq X_n(t_0)$, yield $\sup X_n(t) - X_n(t_0) \to 0$, i.e. (9.2), which implies (9.3) by (9.1).   □

Even in cases where Theorem 16 is formally applicable, it may still be disappointing because the limit in (9.3) is degenerate. For example, if we try to combine Theorem 16 with Theorem 12 for induced subgraph counts (see Section 10), then $a(H;t) \neq 0$ only for $H = K_2$, and $\mathrm{E}\,X_n(t) = \alpha_n f(t)$ exactly, whence Proposition 4.2 yields $\hat{\psi}_n(K_2; t_0) = (n(n-1))^{-1} \alpha_n f'(t_0) = 0$, and thus $Z(t_0) = 0$. In such cases, Theorem 15 may give a non-degenerate limit, but that is not always the case. The following theorem yields further results in a typical case. Note that cancellations similar to those in Theorems 7 and 8 appear.

THEOREM 17. *Suppose that ψ is a graph statistic with finite decomposition such that, for all n,*

$$\hat{\psi}_n(\emptyset; t) = f(t), \tag{9.6}$$

where $f(t)$ has a unique maximum on $[0,1]$ at $t_0 \in (0,1)$ and $f''(t_0) < 0$. Assume further that

$$n^{v(H)} \hat{\psi}_n(H; t_0) \to a(H) \qquad \text{as } n \to \infty, \tag{9.7}$$

for every $H \in \mathcal{U}^0$.

Let

$$\sigma_k^2 = \sum_{H \in \mathcal{U}_k^c} \mathrm{aut}(H)(t_0(1-t_0))^{e(H)} a(H)^2, \qquad k = 3, 4, 5.$$

Then, with $Y_n = \max_t \psi(G_n(t))$,

(i) $n^{3/2}(Y_n - f(t_0)) \xrightarrow{d} N(0, \sigma_3^2)$;

(ii) *if* $\hat{\psi}_n(H; t_0) = 0$ *for* $H \in \mathcal{U}_3^0$, *then*

$$n^2(Y_n - f(t_0)) \xrightarrow{d} N\big(t_0(1-t_0)|f''(t_0)|, \sigma_4^2\big);$$

(iii) *if further* $\hat{\psi}_n(H; t_0) = 0$ *for* $H \in \mathcal{U}_4^c$, *then*

$$n^{5/2}\big(Y_n - (f(t_0) + n^{-2}t_0(1-t_0)|f''(t_0)|)\big) \xrightarrow{d} N\big(0, \sigma_5^2 + 8(f''(t_0))^2(t_0(1-t_0))^2\big).$$

Proof. We decompose $X_n(t) = \psi(G_n(t))$ as

$$X_n(t) = f(t) + \sum_{j=2}^{8} X_{nj}(t), \tag{9.8}$$

where $X_{nj}(t) = \psi^{\mathcal{H}_j}(G_n(t))$, with

$$\mathcal{H}_2 = \mathcal{U}_2^c = \{K_2\}, \qquad\qquad \mathcal{H}_3 = \mathcal{U}_3^c = \{P_2, K_3\},$$
$$\mathcal{H}_4 = \mathcal{U}_4^c, \qquad\qquad\qquad \mathcal{H}_5 = \mathcal{U}_5^c,$$
$$\mathcal{H}_6 = \{K_2 + K_2\}, \qquad\qquad \mathcal{H}_7 = \{P_2 + K_2, K_3 + K_2\},$$
$$\mathcal{H}_8 = \{H \in \mathcal{U}^0 : v(H) \geq 6\}.$$

By (9.7), $\hat{\psi}_n(H; t_0) = O(n^{-v(H)})$ for every H. Hence Proposition 4.3 yields $\frac{d^k}{dt^k}\hat{\psi}_n(H; t_0) = O(n^{-v(H)})$ for $k \geq 1$, and thus, since $\hat{\psi}_n(H; t)$ is a polynomial by Proposition 4.5,

$$\hat{\psi}_n(H; t) = O(n^{-v(H)}), \tag{9.9}$$

$$\frac{d}{dt}\hat{\psi}_n(H; t) = O(n^{-v(H)}), \tag{9.10}$$

uniformly for $t \in [0, 1]$. Using Lemma 3.2 we obtain

$$X_{n2}(t) = O_p(n^{-1}), \qquad X_{n3}(t) = O_p(n^{-3/2}),$$
$$X_{n4}(t) = O_p(n^{-2}), \qquad X_{n5}(t) = O_p(n^{-5/2}),$$
$$X_{n6}(t) = O_p(n^{-2}), \qquad X_{n7}(t) = O_p(n^{-5/2}),$$
$$X_{n8}(t) = O_p(n^{-3}),$$

where the O_p here and below are uniform in $t \in [0, 1]$. (Thus, for example, the sequence $n \sup_t |X_{n2}(t)|$ is bounded in probability (i.e., tight). Equivalently, $n \sup_t |X_{n2}(t)|/\omega(n) \xrightarrow{P} 0$ for every $\omega(n) \to \infty$.)

Since $f'(t_0) = 0$, Proposition 4.2 implies $\hat{\psi}_n(K_2; t_0) = 0$, and (9.10) yields $\hat{\psi}_n(K_2; t) = O(|t - t_0|n^{-2})$ and thus

$$X_{n2}(t) = O_p(|t - t_0|n^{-1}). \tag{9.11}$$

Collecting the terms we obtain

$$X_n(t) = f(t) + X_{n3}(t) + O_p(|t - t_0|n^{-1} + n^{-2}). \tag{9.12}$$

Since the assumptions imply $f(t) \leq f(t_0) - c(t - t_0)^2$ for some $c > 0$, we obtain easily, similarly to the proof of Theorem 16, for $0 < \varepsilon < 1/4$,

$$P\left(\sup_{|t-t_0|>n^{-1/2-\varepsilon}} X_n(t) > X_n(t_0)\right) \to 0 \tag{9.13}$$

and

$$Y_n = f(t_0) + X_{n3}(t_0) + o_p(n^{-3/2}). \tag{9.14}$$

Part (i) now follows by (9.7) and Theorem 1.

In case (ii), (9.10) and the assumption yield $\hat{\psi}_n(P_2; t) = O(|t - t_0|n^{-3})$ and $\hat{\psi}_n(K_3; t) = O(|t - t_0|n^{-3})$, and thus

$$X_{n3}(t) = O_p(|t - t_0|n^{-3/2}). \tag{9.15}$$

Furthermore, by Proposition 4.2,

$$\hat{\psi}_n(K_2 + K_2; t) = \frac{1}{2(n)_4}f''(t) - \frac{1}{n-3}\hat{\psi}_n(P_2; t)$$

$$= \frac{1}{2}n^{-4}f''(t) + O(n^{-5} + |t - t_0|n^{-4}), \tag{9.16}$$

and, by (3.36),

$$S_n(K_2;t)^2 = S_n(K_2 + K_2;t) + 4S_n(P_2;t) + 2(1 - 2t)S_n(K_2;t)$$
$$+ 2n(n - 1)t(1 - t)$$
$$= S_n(K_2 + K_2;t) + 2n^2t(1 - t) + O_p(n^{3/2}). \qquad (9.17)$$

Hence,

$$X_{n6}(t) = \frac{1}{2}n^{-4}f''(t)(S_n(K_2;t))^2 - n^{-2}f''(t)t(1 - t) + O_p(n^{-5/2} + |t - t_0|n^{-2}). \qquad (9.18)$$

By Proposition 4.2,

$$X_{n2}(t) = \hat{\psi}_n(K_2;t)S_n(K_2;t) = \frac{f'(t)}{n(n - 1)}S_n(K_2;t), \qquad (9.19)$$

and we obtain

$$X_n(t) = f(t) + f'(t)\frac{1}{n(n - 1)}S_n(K_2;t) + \frac{1}{2}f''(t)(\frac{1}{n(n - 1)}S_n(K_2;t))^2$$
$$- n^{-2}f''(t)t(1 - t) + X_{n4}(t) + O_p(n^{-5/2} + |t - t_0|n^{-3/2})$$
$$= f\left(t + \frac{1}{n(n - 1)}S_n(K_2;t)\right) - n^{-2}f''(t)t(1 - t) + X_{n4}(t)$$
$$+ O_p(n^{-5/2} + |t - t_0|n^{-3/2}). \qquad (9.20)$$

We put

$$\tilde{t} = t + \frac{1}{n(n - 1)}S_n(K_2;t) = e(G_n(t))/\binom{n}{2}, \qquad (9.21)$$

and see that if t equals the order statistic $T_{(t_0\binom{n}{2})}$, then

$$\tilde{t} = [t_0\binom{n}{2}]/\binom{n}{2} = t_0 + O(n^{-2})$$

and $f(\tilde{t}) = f(t_0) + O(n^{-4})$. Using (9.13) we easily obtain

$$Y_n = f(t_0) - n^{-2}f''(t_0)t_0(1 - t_0) + X_{n4}(t_0) + o_p(n^{-2}), \qquad (9.22)$$

and the result follows.

Finally, in case (iii), we obtain from Proposition 4.3,

$$\frac{d}{dt}\hat{\psi}(P_2;t) = (n - 3)(n - 4)\hat{\psi}_n(P_2 + K_2;t) + O(n^{-3}|t - t_0|),$$

which gives

$$\hat{\psi}(P_2;t) = (n - 3)(n - 4)(t - t_0)\hat{\psi}_n(P_2 + K_2;t) + O(n^{-3}|t - t_0|^2)$$

and

$$\hat{\psi}_n(P_2;t)S_n(P_2;t) = n(n-1)\hat{\psi}(P_2+K_2;t)S_n(P_2;t)(t-t_0)$$
$$+ O_p(|t-t_0|^2 n^{-3/2} + |t-t_0| n^{-5/2}).$$

By (3.36),
$$S_n(P_2+K_2;t) = S_n(P_2;t)S_n(K_2;t) + O_p(n^2), \qquad (9.23)$$

and thus

$$\hat{\psi}_n(P_2;t)S_n(P_2;t) + \hat{\psi}_n(P_2+K_2;t)S_n(P_2+K_2;t)$$
$$= n(n-1)\hat{\psi}_n(P_2+K_2;t)S_n(P_2;t)\big(t-t_0+S_n(K_2;t)/n(n-1)\big)$$
$$+ O_p(|t-t_0|^2 n^{-3/2} + n^{-3}).$$

The same holds with P_2 replaced by K_3, and together these yield

$$X_{n3}(t) + X_{n7}(t) = O_p(n^{-3/2}|\tilde{t}-t_0| + |t-t_0|^2 n^{-3/2} + n^{-3}). \qquad (9.24)$$

Furthermore, the assumption and (9.10) imply

$$X_{n4}(t) = O_p(|t-t_0| n^{-2}), \qquad (9.25)$$

while (9.18) may be sharpened to

$$X_{n6}(t) = \frac{1}{2}n^{-4}f''(t)(S_n(K_2;t))^2 - 2n^{-4}f''(t)S_n(P_2;t)$$
$$- n^{-2}f''(t)t(1-t) + O_p(n^{-3} + |t-t_0| n^{-2}). \qquad (9.26)$$

We thus obtain, similarly to (9.20),

$$X_n(t) = f(\tilde{t}) - n^{-2}f''(t_0)t_0(1-t_0) - 2n^{-4}f''(t)S_n(P_2;t) + X_{n5}(t)$$
$$+ O_p(n^{-3} + |\tilde{t}-t_0| n^{-3/2} + |t-t_0|^2 n^{-3/2} + |t-t_0| n^{-2}),$$

which, by arguments as above, leads to

$$Y_n = f(t_0) - n^{-2}f''(t_0)t_0(1-t_0) - 2n^{-4}f''(t_0)S_n(P_2;t_0) + X_{n5}(t_0) + o_p(n^{-5/2}), \qquad (9.27)$$

and the result follows by Theorem 1. $\qquad\square$

REMARK 9.1. It follows from Theorems 7 and 8 that, in all three cases in Theorem 17, the asymptotics for $\psi(G_{n,m_n})$ with $m_n = t_0\binom{n}{2} + o(n^{3/4})$ are the same as for Y_n. Hence, the maximum of ψ over the evolution is essentially attained when there are $[t_0\binom{n}{2}]$ edges, in the sense that the difference between the maximum and the value at that instance is negligible compared to the random fluctuations. (Our methods are not refined enough to yield precise information about that difference or about the exact number of edges when the maximum is attained.) Note that the asymptotics for $\psi(G_{n,t_0})$ are the same in case (i), but not in cases (ii) and (iii).

III. EXAMPLES

10. Subgraph counts

In this section we apply the general methods to subgraph counts. We begin by considering all (weak) subgraphs, and treat later the corresponding problems for induced subgraphs. For the case p_n constant, see also [10] and [15].

We let in this section $\widetilde{X}$ denote the standardization $(X - \mathrm{E}\, X)/(\mathrm{Var}\, X)^{1/2}$ of a random variable X (assuming $0 < \mathrm{Var}\, X < \infty$).

Let G be a fixed (unlabelled) graph, with $e(G) > 0$ to avoid trivialities, and let $X_G(H) = \mathrm{sub}(G, H)$ be the number of subgraphs of H that are isomorphic to G. Thus,

$$X_G(G_n(t)) = \frac{1}{\mathrm{aut}(G)} \sum_{G_1 \cong G} \prod_{e \in G_1} I_e(t),$$

and if we substitute $I_e(t) = I'_e(t) + t$ $(0 \le t \le 1)$ and expand as in the proof of Proposition 4.1, we obtain the orthogonal decomposition (4.1), with only $H \subseteq G$ appearing. Hence X_G has a finite decomposition and the support of X_G is the set of all subgraphs of G without isolated vertices. Furthermore, it follows from this argument that, for such H,

$$\widehat{X}_{G,n}(H; t) \asymp n^{v(G)-v(H)} t^{e(G)-e(H)} \tag{10.1}$$

and thus

$$\widehat{X}^*_{G,n}(H; t) \asymp n^{v(G)-v(H)/2} t^{e(G)-e(H)/2}. \tag{10.2}$$

Now consider a sequence p_n, and let $X_{n,p_n} = X_G(G_n(p_n))$. If $p_n \to p$, $0 < p < 1$, then $\widehat{X}_{G,n}(H; p_n) \asymp n^{v(G)-v(H)/2}$, and by Proposition 4.6 (and Remark 4.3), X_G is dominated by the $H \ne \emptyset$ with smallest $v(H)$, i.e. by $\{K_2\}$. In particular, by Theorem 3,

$$\widetilde{X}_{n,p_n} \xrightarrow{d} N(0,1).$$

Theorem 10 yields moment convergence, cf. Remark 7.3.

If $p_n \to 0$, then, by (10.1), X_G is dominated by the set of $H \ne \emptyset$ for which $n^{v(H)} p_n^{e(H)}$ is of smallest order (more precisely, by the set of $H \ne \emptyset$ for which $\liminf(n^{v(H)} p_n^{e(H)} / \min_{F \subseteq G} n^{v(F)} p_n^{e(F)}) < \infty$). These H are called *leading overlaps* in [14]. The set of them depends on p_n, and we let the *spectrum* of G be the set of H that are leading overlaps for some sequences p_n. It is shown in [14] that the spectrum can be described as follows. Let $\Omega = \{(v(H), e(H)) : H \subseteq G, v(H) \ge 2\}$. Then H belongs to the spectrum of G if and only if $(v(H), e(H))$ belongs to the top part of the boundary of the convex hull of Ω. Moreover, the range of p_n for which such an H is a leading overlap is described by the slopes of the boundary of this convex hull to the left and right of $(v(H), e(H))$.

If we impose the condition $np_n^{m(G)} \to \infty$, or, equivalently, $n^{v(H)}p^{e(H)} \to \infty$ for all non-empty $H \subseteq G$ (see Remark 3.1), then all leading overlaps are connected, because $H = H_1 + H_2$ with $H_1, H_2 \neq \emptyset$ implies

$$n^{v(H)}p^{e(H)} = n^{v(H_1)}p^{e(H_1)}n^{v(H_2)}p^{e(H_2)} \gg n^{v(H_2)}p^{e(H_2)}. \tag{10.3}$$

Since also $m(H) \leq m(G)$ for $H \subseteq G$, Theorem 3 yields $\widetilde{X}_{n,p_n} \xrightarrow{d} N(0,1)$. Moment convergence follows by Theorem 10, since (7.8) holds for any $H \subseteq G$.

Finally, if $p_n \to 1$, we perform a time reversal as in Proposition 4.11. Thus, with $q_n = 1 - p_n \to 0$, $X_G(G_{n,p_n}) \stackrel{d}{=} \overline{X}_G(G_{n,q_n})$ and, for $H \subseteq G$,

$$|\widehat{\overline{X}}_{G,n}(H; q_n)| = |\widehat{X}_{G,n}(H; p_n)| \asymp n^{v(G)-v(H)}. \tag{10.4}$$

Thus

$$|\widehat{\overline{X}}^{*}_{G,n}(H; q_n)| \asymp n^{v(G)-v(H)/2}q_n^{e(H)/2}, \tag{10.5}$$

and consequently $\overline{X}_G(G_{n,q_n})$ is dominated by K_2. Theorem 3 yields that if $n^2 q_n \to \infty$, then $\widetilde{X}_{n,p_n} \xrightarrow{d} N(0,1)$. This time some work is needed in order to apply Theorem 10. We note that (7.9) may be written, using (10.5) and (7.1) and assuming $H \subseteq G$,

$$n^{v(G)-v(H)/2}q_n^{e(H)/2} = o(\beta_n n^{v(F)(1/2-\varepsilon)}q_n^{e(F)(1/2-\varepsilon)}) \tag{10.6}$$

for every $F \subseteq H$, with $\beta_n \asymp \widehat{\overline{X}}^{*}_{G,n}(K_2; q_n)$. It is easily seen that if H' is obtained by removing an edge from H, together with any end point that otherwise would become isolated, $F' = F \cap H'$, and (10.6) holds with O for H' and F', then (10.6) holds. By induction, this proves (10.6), and thus (7.9) (with q_n), for every H with $e(H) \geq 2$. Since (7.8) holds for K_2, Theorem 10 yields convergence of all moments.

Combining these results, using Proposition 2.5, we obtain the following result by Ruciński [24].

THEOREM 18. *If* $np_n^{m(G)} \to \infty$ *and* $n^2(1 - p_n) \to \infty$, *then* $\widetilde{X}_{n,p_n} \xrightarrow{d} N(0,1)$ *with convergence of all moments.* $\square$

(The conditions are necessary too, see [24].)

Next we consider $G_{n,m}$ and let $X_{n,m_n} = X_G(G_{n,m_n})$. We apply Theorem 7 and now have to find the set of $H \neq K_2$ for which $n^{v(H)}p_n^{e(H)}$ are of smallest order, where $p_n = m_n/\binom{n}{2}$. This can be done as above, replacing the set Ω by $\Omega' = \{(v(H), e(H)) : H \subseteq G, v(H) \geq 3\}$.

If G is not a sum of isolated vertices and edges, and $np_n^{m(G)} \to \infty$, then $P_2 \subseteq G$ and it is easily seen, using (10.3) again and the obvious fact $n^{v(2K_2)}p_n^{e(2K_2)} \gg n^{v(P_2)}p_n^{e(P_2)}$, that the extremal H are connected. Hence X_{n,m_n} is dominated by a finite family of connected graphs, and Theorems 7 and 11 apply with $a(H) = 0$

whenever H is disconnected, which gives convergence to a normal limit. This is true also when $p_n \to 1$, in which case X_{n,m_n} is dominated by $\{P_2\}$, and the conditions in Theorem 11 may be verified as above.

In the exceptional case $G = iK_1 + jK_2$, we assume $j \geq 2$ since otherwise X_{n,m_n} is constant. Then $\widehat{X}_{G,n}(H;t) \neq 0$ only when $H = kK_2$, $k \leq j$, and Theorem 7 is not directly applicable. Theorem 8 applies provided $np_n \to \infty$, but not for smaller p_n. In general, we may argue as in Theorem 8, subtracting $\binom{e(H)}{j}\binom{v(H)-2j}{i}$ from $X_G(H)$. It then follows by straightforward estimates, which we leave to the reader, that X_{n,m_n} is dominated by $\{P_2\}$, cf. [10], and we may apply Theorem 7 to the modified statistic. Consequently, asymptotic normality holds also for such G, provided $n^3 p_n^2 \to \infty$ and $n^3(1 - p_n)^2 \to \infty$. Furthermore, again Theorem 11 implies that then all moments converge, since (7.17) holds at least for $H = kK_2$ and P_2, and it may be shown by induction as above that (7.18) holds for all other H.

Consequently, combining the two cases:

THEOREM 19. *If $e(G) > 1$, $p_n = m_n/\binom{n}{2}$, $np_n^{m(G)} \to \infty$, $n^3 p_n^2 \to \infty$ (redundant when $P_2 \subseteq G$), and $n^3(1 - p_n)^2 \to \infty$, then*

$$\widetilde{X}_{n,m_n} \xrightarrow{d} N(0,1), \tag{10.7}$$

with convergence of all moments.   □

We next note that by Remarks 5.1 and 6.4, joint convergence to a multivariate normal distribution holds in Theorems 18 and 19 for any family of subgraph counts $X_{G(i)}$ such that the conditions on p_n are satisfied for each $G(i)$. If $p_n = p$ is fixed, then every $X_{G(i)}(G_{n,p})$ is dominated by K_2, and thus the limit distribution is singular and concentrated on a 1-dimensional subspace, i.e. the subgraph counts are asymptotically linear functions of each other. The same is true for G_{n,m_n} if $p_n = m_n/\binom{n}{2} \to 1$ (P_2 dominates), while $p_n \to p \in (0,1)$ yields (in general) a two-dimensional limit distribution (P_2 and K_3 dominate). If $p_n \to 0$, we can obtain non-singular limit distributions of higher dimension.

We next consider induced subgraph counts. Let $Y_G(H)$ be the number of induced subgraphs of H that are isomorphic to G. We see, as before, that Y_G has a finite decomposition, but now the spectrum consists of all graphs $H \in \mathcal{U}^0$ with $v(H) \leq v(G)$.

Let us first consider the case $p_n \to 0$. It is then easily seen that if $H \subseteq G$, then $\widehat{Y}_{G,n}(H;p_n) = \widehat{X}_{G,n}(H;p_n)(1+O(p_n)) \sim \widehat{X}_G(H;p_n)$. It is also easy to show that, unless G is discrete, $\widehat{Y}^*_{G,n}(H;p_n) = o(\max_{\emptyset \neq F \subseteq G} \widehat{Y}^*_{G,n}(F;p_n))$ whenever $H \not\subseteq G$. Consequently Y_G is dominated by the same family of subgraphs as X_G, and $Y_{n,p_n} = Y_G(G_{n,p_n})$ is asymptotically normal. In the exceptional case when G is discrete (and $v(G) \geq 2$), X_G is constant, while Y_G is dominated by $\{K_2\}$ and is thus asymptotically normal as soon as $n^2 p_n \to \infty$. In both cases, the conditions of Theorem 10 may be verified by induction on H, which gives the following result.

THEOREM 20. *If $v(G) \geq 2$, $p_n \to 0$, $n^2 p_n \to \infty$ and $np_n^{m(G)} \to \infty$, then $\widetilde{Y}_{n,p_n} \xrightarrow{d} N(0,1)$ with convergence of all moments.*   □

For $Y_{n,m} = Y_G(G_{n,m})$ there is a corresponding analogue of Theorem 19. The only real difficulty is when $G = iK_1 + jK_2$. In this case, we have to assume $v(G) \geq 3$, since otherwise Y_{n,m_n} is constant. We then subtract a function of m_n and n as above and obtain a linear combination of X_{G_α}, with $G_\alpha \supseteq P_2$. Some calculations, which we omit, then show that Y_{m,n_m} is dominated by $\{P_2, P_3\}$ (and by P_2 alone if $e(G) \leq 1$ or $m_n/n \to 0$ or $m_n/n \to \infty$). It is possible (but tedious) to verify the conditions of Theorem 11 in this case too, provided $n^3 p_n^2 \to \infty$. This gives the following result.

THEOREM 21. *If $v(G) \geq 3$, $p_n = m_n/\binom{n}{2} \to 0$, $np_n^{m(G)} \to \infty$ and $n^3 p_n^2 \to \infty$ (redundant when $P_2 \subseteq G$), then $\widetilde{Y}_{n,m_n} \xrightarrow{d} N(0,1)$ with convergence of all moments.* $\qquad\square$

For $p_n \to 1$ we get the same results, since $Y_G(H) = Y_{-G}(-H)$. Note that the condition $np_n^{m(G)} \to \infty$ is replaced by $n(1 - p_n)^{m(-G)} \to \infty$.

Finally, let us consider the case $p_n \to p \in (0, 1)$. Since

$$\mathrm{E}\, Y_G(G_n(t)) = \alpha_n t^{e(G)}(1-t)^{e(-G)}, \tag{10.8}$$

where $\alpha_n = (n)_{v(G)}/\operatorname{aut}(G)$, Proposition 4.2 shows that

$$\widehat{Y}_{G,n}(K_2; t) = \frac{\alpha_n}{n(n-1)}\left(e(G) - \binom{n}{2}t\right) t^{e(G)-1}(1-t)^{e(-G)-1}.$$

Similarly, for every H there is a polynomial π_H such that

$$\widehat{Y}_{G,n}(H; t) = \frac{(n)_{v(G)}}{(n)_{v(H)}}\pi_H(t). \tag{10.9}$$

If $p \neq e(G)/\binom{v(G)}{2}$, then Y_{n,p_n} is dominated by $\{K_2\}$, and is thus asymptotically normal. If, however, $p_n = p = e(G)/\binom{v(G)}{2}$ (G neither discrete nor complete), then $\widehat{Y}_{G,n}(K_2; p) = 0$. In this case, Y_{n,p_n} is dominated by $\{P_2, K_3\}$, unless also $\widehat{Y}_{G,n}(P_2; p) = \widehat{Y}_{G,n}(K_3; p) = 0$.

Graphs G that satisfy these conditions are known as *p-proportional*. For such G and $p_n = p$, Y_G is dominated by $\{H : v(H) = 4\}$. Furthermore, $\operatorname{Var}(Y_{n,p}) \asymp n^{2v(G)-4}$ and, e.g. by (4.10), $\lim_{n\to\infty} \hat{\psi}_n^*(K_2 + K_2; p)/\operatorname{Var}(Y_{n,p})^{1/2} \neq 0$. Hence Theorem 5 (or 4) shows that $Y_{n,p}$ is not asymptotically normal.

More precisely, Theorem 2 yields the following result, see [3] and [10] for further details. See also [11] for more on $\widehat{Y}_{G,n}(H; t)$, and [13], [18] and [16] for constructions of proportional graphs.

THEOREM 22. *Let $0 < p < 1$.*

(i) *Then*

$$n^{1-v(G)}(Y_{n,p} - \mathrm{E}\,Y_{n,p}) \xrightarrow{d} N(0, \sigma_2^2), \tag{10.10}$$

for some $\sigma_2^2 \geq 0$;

(ii) *$\sigma_2^2 = 0$ if and only if $p = e(G)/\binom{v(G)}{2}$, and then*

$$n^{3/2-v(G)}(Y_{n,p} - \mathrm{E}\,Y_{n,p}) \xrightarrow{d} N(0, \sigma_3^2), \tag{10.11}$$

for some $\sigma_3^2 \geq 0$;

(iii) $\sigma_2^2 = \sigma_3^2 = 0$ if and only if G is p-proportional, and then

$$n^{2-v(G)}(Y_{n,p} - \mathrm{E}\,Y_{n,p}) \xrightarrow{d} a(Z_1^2 - 1) + bZ_2, \qquad (10.12)$$

where $Z_1, Z_2 \sim N(0,1)$ are independent, a, b are constants, and $a < 0$. This limit is non-degenerate and not normal.

In every case, all moments converge. The parameters in the limit distributions are given by

$$\sigma_k^2 = \sum_{H \in \mathcal{U}_k^c} \mathrm{aut}(H)(p(1-p))^{e(H)} \pi_H(p)^2, \qquad (10.13)$$

where π_H is as in (10.9),

$$a = 2p(1-p)\pi_{K_2+K_2}(p) = -\frac{1}{\mathrm{aut}(G)}\binom{v(G)}{2}p^{e(G)}(1-p)^{e(-G)} \qquad (10.14)$$

and $b = \sigma_4$, given by (10.13). $\qquad\qquad\qquad\qquad\qquad\qquad\qquad\square$

We say, more generally, that the graph G is $(p, \mathcal{H})\hat{}$-*proportional*, where $0 < p < 1$ and $\mathcal{H} \subseteq \mathcal{U}^0$, if $v(G) \geq \max\{v(H) : H \in \mathcal{H}\}$ and $\hat{Y}_{G,n}(\mathcal{H};p) = 0$ for $H \in \mathcal{H}$ and $n \geq 1$. Some equivalent conditions are given in [11]. (Note that the Fourier transform $\hat{\psi}$ in [11] uses a different normalization than the one in the present paper; they vanish, however, simultaneously.) For example, p-proportional is the same as $(p, \mathcal{U}_2^0 \cup \mathcal{U}_3^0)\hat{}$-proportional, and it follows from the argument above that the asymptotic distribution of $Y_{n,p}$ is a χ^2-distribution if and only if G is $(p, \mathcal{U}_2^0 \cup \mathcal{U}_3^0 \cup \mathcal{U}_4^c)\hat{}$-proportional. (It is not obvious that such graphs, which may be called *superproportional*, exist. An example with 64 vertices and $p = 1/2$ has recently been found by Jan Kärrman [19].)

For $G_{n,m}$, with $m \asymp n^2$, we have the following analogue of Theorem 22.

THEOREM 23. *Let* $0 < p < 1$ *and* $m_n = [\binom{n}{2}p]$, *or, more generally,* $m_n = \binom{n}{2}p + o(n^{1/2})$.

(i) *Then*

$$n^{3/2-v(G)}(Y_{n,m_n} - \mathrm{E}\,Y_{n,m_n}) \xrightarrow{d} N(0, \sigma_3^2), \qquad (10.15)$$

for some $\sigma_3^2 \geq 0$;

(ii) $\sigma_3^2 = 0$ *if and only if* G *is* $(p, \mathcal{U}_3^0)\hat{}$-*proportional, and then*

$$n^{2-v(G)}(Y_{n,m_n} - \mathrm{E}\,Y_{n,m_n}) \xrightarrow{d} N(0, \sigma_4^2), \qquad (10.16)$$

for some $\sigma_4^2 \geq 0$;

(iii) $\sigma_3^2 = \sigma_4^2 = 0$ *if and only if* G *is* $(p, \mathcal{U}_3^0 \cup \mathcal{U}_4^c)\hat{}$-*proportional, and then*

$$n^{5/2-v(G)}(Y_{n,m_n} - \mathrm{E}\,Y_{n,m_n}) \xrightarrow{d} N(0, \sigma_5^2), \qquad (10.17)$$

for some $\sigma_5^2 \geq 0$;

(iv) $\sigma_3^2 = \sigma_4^2 = \sigma_5^2 = 0$ if and only if G is $(p, \mathcal{U}_3^0 \cup \mathcal{U}_4^0 \cup \mathcal{U}_5^c)^\frown$-proportional, and then

$$n^{3-v(G)}(Y_{n,m_n} - \mathrm{E}\, Y_{n,m_n}) \overset{d}{\to} a(Z_1^2 - 1) + b(Z_2^2 - 1) + cZ_3, \qquad (10.18)$$

where $Z_1, Z_2, Z_3 \sim N(0,1)$ are independent, a, b, c are constants, and at least $a \neq 0$. This limit is non-degenerate and not normal.

In every case, all moments converge as well.

Proof. Cases (i) and (ii) follow directly by Theorem 7, and (iii) by Theorem 8. Moment convergence follows by Theorem 11 (and Remark 7.6), noting that (7.17) holds trivially because $p > 0$. (The σ_k^2 are as in (10.13) for $k = 3$ and 4, while σ_5^2 contains an additional term $32(p(1-p))^2 \pi_{K_2+K_2}(p)^2$.)

For (iv), we study Y_{n,p_n} more closely, where $p_n = m_n/\binom{n}{2}$ as usual. We first take care of the term with $S_n(P_2 + K_2; p_n)$, whose coefficient (with the given normalization) is of the order n^{-2}, so that the whole term has variance of the order $n^{-4+5} = n$. (The $S_n(K_3+K_2; p_n)$ term could be handled similarly, but this is not necessary, since the coefficient $n^{3-v(G)}\hat{Y}_{G,n}(K_3 + K_2; t)$ actually vanishes when $t = p$, see [11, Example 3.1], and thus the coefficient is $O(n^{-2}|p_n - p|) = o(n^{-5/2})$ when $t = p_n$.) We argue as in the proof of Theorem 8, subtracting from Y_G a function $\psi(H) = c(e(H) - m_n)\psi_1(H)$, where c is a suitable constant, and the statistic ψ_1 satisfies $\psi_1(G_n(p_n)) = S_n(P_2, p_n)$, cf. Remark 4.2. This does not affect Y_{n,m_n}, since $\psi(G_{n,m_n}) = 0$, but affects the decomposition of $Y_G(G_n(p_n))$ by replacing $S_n(P_2 + K_2; p_n)$ by $-4S_n(P_3; p_n) - 2S_n(K_{1,3}; p_n) +$ smaller order terms, where $K_{1,3}$ is a star with 4 vertices. Since the coefficient of these terms (after normalization) is of order n^{-2}, the new terms have variances of order 1; the terms with $S_n(H; p_n)$, $v(H) = 6$, also have variances of (at most) this order, while all other terms are smaller.

We now apply Theorem 11 to the modified Y_G and obtain

$$n^{3-v(G)}(Y_{n,m_n} - \mathrm{E}\, Y_{n,m_n}) \overset{d}{\to} a'(Z_1^2 - 1) + b'Z_1 Z_2 + c'(Z_2^2 - 1) + d'Z_3, \quad (10.19)$$

where Z_3 is a linear combination of $U(H; 1)$ for $H \in \mathcal{U}_6^c \cup \{P_3, K_{1,3}\}$ (from the previous argument), while the quadratic terms come from the disconnected graphs $K_3 + K_3$, $K_3 + P_2$ and $P_2 + P_2$ in $\tilde{\mathcal{U}} \cap \mathcal{U}_6^0$. The simpler expression (10.18) follows by a diagonalization of the quadratic form. That $b' \neq 0$, and thus the limit is non-degenerate and, by Lemma 5.1, non-normal, follows because $\hat{Y}_{G,n}(K_3 + P_2; p) \neq 0$ for any such G and p, which follows from [11, Example 3.2]. □

Kärrman's example [19] shows that there exist a graph G such that case (iii) occurs. We do not, however, know if there exist any graph such that case (iv) happens; thus we do not know whether $Y_G(G_{n,m_n})$ always is asymptotically normal. (We conjecture that such graphs, yielding a non-normal limit, exist.) We note that if case (iv) happens, then $\hat{Y}_G(P_2; p) = \hat{Y}_G(K_2 + K_2; p) = 0$, and thus, by Proposition 4.2, p has to be an inflection point of $\mathrm{E}\, Y_G(G_n(t))$, or, by (10.8), of $t^{e(G)}(1-t)^{e(-G)}$, which by a straightforward calculation gives the necessary condition $p = \frac{e(G)}{M} \pm \frac{1}{M}\sqrt{\frac{e(G)(M-e(G))}{M-1}}$, where $M = \binom{v(G)}{2}$. (This also follows from [11, Theorem 4].) In particular, such graphs are not proportional, so the classes of graphs that yield non-normal limits in Theorems 22 and 23 are disjoint.

Let us finally consider the maximum number of induced copies of G during the evolution. We will see that, in contrast to Theorems 22 and 23, we always obtain an asymptotic normal distribution.

THEOREM 24. *Let* $Y_n = \max_t Y_G(G_n(t))$, *where G is neither complete nor discrete, and let*

$$\gamma_n = e(G)^{e(G)}(M - e(G))^{M-e(G)}M^{-M}(\operatorname{aut} G)^{-1}(n)_{v(G)},$$

where $M = \binom{v(G)}{2}$, *and*

$$\gamma_n' = \gamma_n(1 + Mn^{-2}).$$

(i) *Then*

$$n^{3/2}(Y_n/\gamma_n - 1) \xrightarrow{d} N(0, \sigma_3^2),$$

with $\sigma_3^2 \geq 0$;

(ii) $\sigma_3^2 = 0$ *if and only if G is proportional, and then*

$$n^2(Y_n/\gamma_n' - 1) \xrightarrow{d} N(0, \sigma_4^2)$$

with $\sigma_4^2 \geq 0$;

(iii) $\sigma_3^2 = \sigma_4^2 = 0$ *if and only if G is superproportional, and then*

$$n^{5/2}(Y_n/\gamma_n' - 1) \xrightarrow{d} N(0, \sigma_5'^2),$$

with $\sigma_5'^2 = \sigma_5^2 + 8M^2 > 0$.

Proof. Follows by Theorem 17 and simple computations.   □

Here σ_k^2 are as in Theorem 17. We have chosen slightly different normalizations in the last two theorems, but apart from that, the limit for the maximum is the same as for Y_{n,m_n} with $m_n \approx \frac{e(G)}{M}\binom{n}{2}$, cf. Remark 9.1.

11. Vertex degrees

Let $\psi_d(G)$ be the number of vertices of degree d in G, and let $X_{dn}(t) = \psi_d(G_n(t))$, where $d \geq 0$ is fixed. Simple computations, using $X_{dn}(t) = \sum_1^n I(i$ has degree d in $G_n(t))$, yield

$$\operatorname{E} X_{dn}(t) = n\binom{n-1}{d}t^d(1-t)^{n-1-d} \tag{11.1}$$

and

$$\operatorname{Var} X_{dn}(t) = n\binom{n-1}{d}t^d(1-t)^{n-1-d}\left(1 - \binom{n-1}{d}t^d(1-t)^{n-1-d}\right)$$
$$+ n(n-1)t(1-t)\left(\binom{n-2}{d-1}t^{d-1}(1-t)^{n-1-d} - \binom{n-2}{d}t^d(1-t)^{n-2-d}\right)^2. \tag{11.2}$$

In particular, if $nt \to 0$,

$$\operatorname{Var} X_{dn}(t) \sim \begin{cases} 2n^2 t, & d = 0, 1, \\ n(nt)^d/d!, & d \geq 2; \end{cases} \tag{11.3}$$

if $nt \to \lambda > 0$,

$$n^{-1} \operatorname{Var} X_{dn}(t) \to \frac{\lambda^d}{d!} e^{-\lambda}\left(1 - \frac{\lambda^d}{d!} e^{-\lambda}\right) + \frac{(\lambda - d)^2}{\lambda}\left(\frac{\lambda^d}{d!} e^{-\lambda}\right)^2; \tag{11.4}$$

and if $nt \to \infty$ while $nt^2 \to 0$,

$$\operatorname{Var} X_{dn}(t) \sim n \frac{(nt)^d e^{-nt}}{d!}. \tag{11.5}$$

Summarizing, for fixed d, if $n \geq (d+1) \vee 2$ and $nt^2 \to 0$,

$$\operatorname{Var} X_{dn}(t) \asymp \begin{cases} n(1 \wedge nt)e^{-nt}, & d = 0, \\ n(nt)^d e^{-nt}, & d \geq 1. \end{cases} \tag{11.6}$$

We have, summing over ordered sets of distinct indices,

$$X_{dn}(t) = \sum_i \frac{1}{d!} \sum_{j_1 \ldots j_d} \prod_{r=1}^d I_{ij_r}(t) \prod_{k \neq i, j_r} (1 - I_{ik}(t))$$

$$= \sum_i \sum_{j_1 \ldots j_d} \frac{1}{d!} \prod_{r=1}^d (I'_{ij_r}(t) + t) \prod_{k \neq i, j_r} (1 - t - I'_{ik}(t))$$

$$= \sum_i \sum_{l,m \geq 0} \frac{1}{l!(d-l)!m!} \sum_{j_1 \ldots j_l} \sum_{k_1 \ldots k_m} (n - 1 - l - m)_{d-l} t^{d-l}$$

$$\cdot \prod_{r=1}^l I'_{ij_r}(t) \prod_{s=1}^m (-I'_{ik_s}(t))(1-t)^{n-1-d-m}$$

$$= \sum_{l,m \geq 0} \frac{(-1)^m}{l!(d-l)!m!}(n - 1 - l - m)_{d-l} t^{d-l}(1-t)^{n-1-d-m} S_n(K_{1,l+m}; t),$$

where $K_{p,q}$ denotes the complete bipartite graph. Hence the support of ψ_d is the set $\emptyset \cup \{K_{1,k}\}_{k=1}^\infty$ of the empty graph and all stars (note that $K_{1,1} = K_2$ and $K_{1,2} = P_2$), and

$$\hat{\psi}_{dn}(K_{1,k}; t) = \sum_{l=0}^{k \wedge d} \frac{(-1)^{k-l}}{l!(d-l)!(k-l)!}(n - 1 - k)_{d-l} t^{d-l}(1-t)^{n-1-d-k+l}. \tag{11.7}$$

In particular, if $n \to \infty$ and $nt \to \lambda$,

$$\hat{\psi}_{dn}(K_{1,k}; t) \to \sum_{l=0}^{k \wedge d} \frac{(-1)^{k-l}}{l!(d-l)!(k-l)!}\lambda^{d-l} e^{-\lambda}$$

$$= \frac{(-1)^{k+d}}{k!} \frac{\lambda^d e^{-\lambda}}{d!} \sum_{l=0}^d \binom{d}{l}(-1)^{d-l}\lambda^{-l}(k)_l$$

$$= \frac{(-1)^{k+d}}{k!} \frac{\lambda^d e^{-\lambda}}{d!} C_d(\lambda; k), \tag{11.8}$$

where $C_d(\lambda; k)$ is the dth Charlier polynomial.

REMARK 11.1. The Charlier polynomials are the orthogonal polynomials for the Po(λ) distribution. The orthogonality (and normalization)

$$\sum_{k=0}^{\infty} C_i(\lambda; k) C_j(\lambda; k) \frac{\lambda^k}{k!} e^{-\lambda} = \delta_{ij} \lambda^{-i} i! \tag{11.9}$$

is easily proved using the generating function

$$\sum_{i=0}^{\infty} C_i(\lambda; k) \frac{z^i}{i!} = e^{-z} (1 + \frac{z}{\lambda})^k. \tag{11.10}$$

We will later see that it is not a coincidence that these orthogonal polynomials appear here.

By (11.7),

$$|\hat{\psi}_{dn}(K_{1,k}; t)| \leq \frac{1}{k!} \sum_{l=0}^{d} (k)_l \binom{d}{l} (nt)^{d-l} \leq \frac{k^d}{k!} (1 + nt)^d, \tag{11.11}$$

and thus, if $A = \sup np_n < \infty$,

$$|\hat{\psi}_{dn}^*(K_{1,k}; t)| \leq n^{1/2} A^{k/2} \frac{k^d}{k!} (1 + A)^d.$$

Since

$$\sum_{k=2}^{\infty} \mathrm{aut}(K_{1,k}) \left(A^{k/2} \frac{k^d}{k!} (1 + A)^d \right)^2 = \sum_{k=2}^{\infty} \frac{1}{k!} k^{2d} A^k (1 + A)^{2d} < \infty,$$

Proposition 4.8 and (11.4) imply that $X_{dn}(p_n)$ is almost finitely dominated by $\{K_{1,k}\}$ when $np_n \to \lambda > 0$. Alternatively, this may be proved using Proposition 4.7 (with $\beta_n = n^{1/2}$), since calculations with (11.4) and (11.8) show that in this case (4.25) follows from (11.9).

It follows similarly that the same is true when $np_n \to 0$; moreover, then $X_{dn}(p_n)$ is dominated by the one-element set $\{K_{1,d}\}$ if $d \geq 1$, and by $\{K_{1,1}\} = \{K_2\}$ when $d = 0$.

If $np_n \to \infty$, $X_{dn}(p_n)$ is not almost finitely dominated; in fact, the set in (4.22) and (4.23) is empty.

Barbour, Karoński and Ruciński [3] proved that $X_{dn}(p_n)$ is asymptotically normal if (and only if) $\mathrm{E}\, X_{dn}(p_n) \to \infty$ and, if $d = 0$, $n^2 p_n \to \infty$. (This is equivalent to $n(np_n)^{d \vee 1} \to \infty$ and $np_n - \log n - d \log \log n \to -\infty$.) The methods of the present paper are not adapted to the case $np_n \to \infty$, when ψ_d is not almost finitely dominated, but otherwise Theorem 3 immediately gives the following partial result.

THEOREM 25. If $np_n = O(1)$ and $n(np_n)^{d\vee 1} \to \infty$, then

$$(X_{dn}(p_n) - \mathrm{E}\, X_{dn}(p_n))/(\mathrm{Var}\, X_{dn}(p_n))^{1/2} \xrightarrow{d} N(0,1). \qquad (11.12)$$

$\square$

In the remaining case, $np_n \to \infty$, Theorem 6 is applicable, using arguments similar to those in [2] to establish (5.14), but it seems much simpler to use Stein's method as in [3] or Stein's method for Poisson convergence as in [17] and [1, Theorem 5.F].

We next observe that the drift of $X_{dn}(t)$ can be computed by Proposition 2.8 to be

$$\xi_{dn}(t) = \frac{1}{1-t}((n-d)X_{d-1,n}(t) - (n-d-1)X_{d,n}(t)). \qquad (11.13)$$

(If $d = 0$, we let $X_{-1,n}(t) = 0$.) It follows that if $n \geq d+3$ and $0 < t < 1$, then

$$\mathrm{Var}(\xi_{dn}(t)) \leq 2n^2(1-t)^{-2}\mathrm{Var}\, X_{d-1,n} + 2n^2(1-t)^{-2}\mathrm{Var}\, X_{d,n}$$
$$= O(n^3((nt)^{d-1} + (nt)^d)e^{-nt}). \qquad (11.14)$$

Note that this estimate, (11.6) and Proposition 4.10 yield yet another proof of the almost finite domination of ψ_d when $np_n = O(1)$.

We now can obtain limits for $G_{n,m}$.

THEOREM 26. Let $m_n \to \infty$ and define $p_n = m_n/\binom{n}{2}$, $\alpha_n = \mathrm{E}\,\psi_d(G_{n,p_n})$ and $\beta_n^2 = \mathrm{Var}(\psi_d(G_{n,p_n}))$.

(i) If $m_n/n \to 0$ and $n(m_n/n)^{d\vee 2} \to \infty$, then

$$(\psi_d(G_{n,m_n}) - \alpha_n)/\beta_n' \to N(0,1), \qquad (11.15)$$

where

$$\beta_n' = \begin{cases} (\frac{1}{2}n^3p_n^2)^{1/2} \sim (2m_n^2/n)^{1/2}, & d = 0, \\ (2n^3p_n^2)^{1/2} \sim (8m_n^2/n)^{1/2}, & d = 1, \\ \beta_n, & d \geq 2. \end{cases}$$

(ii) If $2m_n/n \to \lambda$, with $0 < \lambda < \infty$, then

$$n^{-1/2}(\psi_d(G_{n,m_n}) - n\tfrac{(2m_n/n)^d}{d!}e^{-2m_n/n}) \to N(0,\sigma^2), \qquad (11.16)$$

with

$$\sigma^2 = \frac{\lambda^d}{d!}e^{-\lambda}(1 - \frac{\lambda^d}{d!}e^{-\lambda}) - \frac{(\lambda - d)^2}{\lambda}(\frac{\lambda^d}{d!}e^{-\lambda})^2 > 0.$$

(iii) If $m_n/n \to \infty$ and $2m_n/n - \log n - d\log\log n \to -\infty$, then

$$(\psi_d(G_{n,m_n}) - \alpha_n)/\beta_n \to N(0,1). \qquad (11.17)$$

Proof. We use Theorem 9. In all cases, (6.42) (with $\gamma = 1$) follows immediately from (11.3)–(11.5), since $nt_n = np_n + o(1)$ and $t_n \sim p_n$. Furthermore, by (11.14), treating the case $d = 0$ separately,

$$\mathrm{Var}\big(\xi_{dn}(t_n)\big) = O\big(n^2(1 + \tfrac{1}{nt_n})\,\mathrm{Var}\,X_{dn}(t_n)\big) = O\big(n^2\beta_n^2\tfrac{1}{p_n}(p_n + \tfrac{1}{n})\big),$$

which gives (6.43).

In case (i), (11.7) yields $\hat{\psi}_{dn}(K_{1,1}; p_n) \asymp (np_n)^{d-1}$ and thus $\hat{\psi}_{dn}^*(K_2; p_n) \asymp n^d p_n^{d-1/2}$, while (11.6) yields (for $d \geq 1$) $\beta_n \asymp n^{(d+1)/2}p_n^{d/2}$, and (6.41) follows provided $d \geq 2$. Consequently, Theorems 9(i) and 25 then yield (11.15); we may alternatively use Theorem 9(iv), with $a(K_2) = 0$. We treat the cases $d = 0$ or 1 later.

Similarly, in case (iii), $\hat{\psi}_{dn}(K_2; p_n) \asymp (np_n)^d e^{-np_n}$ and (6.41) follows, whence Theorem 9(i) shows that the result follows from the result in [3] and [17] discussed above.

In case (ii), it is convenient to use instead $\beta_n = n^{1/2}$, which does not affect (6.42) and (6.43) because $\mathrm{Var}\,X_n(p_n) \asymp n$, except that now γ is given by the right hand side of (11.4). Moreover, (11.8) yields $a(K_2) = \lim \hat{\psi}_n^*(K_2; p_n)/n^{1/2} = \lambda^{1/2}\frac{\lambda^d e^{-\lambda}}{d!}(\frac{d}{\lambda} - 1)$. Thus the result follows by Theorem 9(iv) and (11.1). We may also use Theorem 9(iii), since (6.46) is trivial. That $\sigma^2 > 0$ follows then easily using (6.52), since $a(H) = 0$ for all $H \neq K_2$ by (11.8) would imply $C_d(\lambda; k) = 0$ for $k = 2, 3, \ldots$, a contradiction because $C_d(\lambda; k)$ is a non-zero polynomial in k.

Finally, if $np_n \to 0$ and $d \leq 1$, we use Theorem 9(ii). Again (6.46) is trivial, while (6.48) follows by

$$\mathrm{Var}(\eta_n(t_n)) \leq \mathrm{Var}(\xi_n(t_n)) = O(n^2 \cdot n^2 t_n) = o(n^2(\beta_n')^2/t_n).$$

We decompose Y_n, obtaining, with $\mathcal{H}_k = \{K_{1,j} : j \geq k\}$,

$$Y_n(t) = X_n^{\mathcal{H}_2}(t) = X_n^{\{K_{1,2}\}}(t) + X_n^{\mathcal{H}_3}(t).$$

It follows easily from (11.11) and Proposition 1.1 that

$$\mathrm{Var}(X_n^{\mathcal{H}_3}(t_n)) = O(n^4 t_n^3) = o((\beta_n')^2), \tag{11.18}$$

while (11.7) yields

$$\hat{\psi}_{dn}(K_{1,2}; t_n) \to \begin{cases} \frac{1}{2}, & d = 0, \\ -1, & d = 1, \end{cases} \tag{11.19}$$

and thus

$$\mathrm{Var}(X_n^{\{K_{1,2}\}}(t_n)) \sim (\beta_n')^2,$$

which together with (11.18) gives (6.47). Further, $(2n^3p_n^2)^{-1/2}S_n(K_{1,2}; p_n) \xrightarrow{d} N(0,1)$ by Theorem 1, which yields $X_n^{\{K_{1,2}\}}(p_n)/\beta_n' \xrightarrow{d} N(0,1)$ and $Y_n(p_n)/\beta_n' \xrightarrow{d} N(0,1)$ by (11.19) and (11.18). The conclusion now follows by Theorem 9(ii). $\qquad\square$

REMARK 11.2. We have proved that $\psi_{dn}(G_{n,m_n})$ with a certain normalization converges to a standard normal distribution. We expect the same to be true with the natural normalization with the mean and variance, but that has not yet been proved.

Next we consider functional limits.

THEOREM 27. *If $d \geq 0$, then*

$$n^{-1/2}\left(X_{dn}(\tfrac{t}{n}) - n\tfrac{t^d}{d!}e^{-t}\right) \xrightarrow{d} Z_d(t)$$

in $D([0, \infty])$, where $Z_d(t)$ is a continuous Gaussian process with $\mathrm{E}\, Z_d(t) = 0$ and

$$\mathrm{Cov}(Z_d(s), Z_d(t)) = \frac{s^d}{d!}e^{-t} - \frac{s^d}{d!}e^{-s}\frac{t^d}{d!}e^{-t} + (s-d)\frac{s^d}{d!}e^{-s}(t-d)\frac{t^{d-1}}{d!}e^{-t},$$

for $0 \leq s \leq t < \infty$.

Proof. Theorem 14(ii) applies with $p_n = 1/n$ and $\beta_n = n^{1/2}$, choosing $\varphi(t) = C(1 + t^{d/2})e^{-t/2}$ for a large constant C, which gives (8.9) by (11.14). By (11.8),

$$a_{dn}(K_{1,k}; t)/\beta_n = \hat{\psi}_{dn}(K_{1,k}; t/n) \to \frac{(-1)^{k+d}}{k!}\frac{t^d e^{-t}}{d!}C_d(t; k),$$

which yields the representation

$$Z_d(t) = \sum_{k=1}^{\infty} \frac{(-1)^{k+d}}{k!}\frac{t^d e^{-t}}{d!}C_d(t; k)U(K_{1,k}; t). \tag{11.20}$$

The covariance formula follows either from this and a calculation with the Charlier polynomials or by taking the limit of $n^{-1}\,\mathrm{Cov}(X_{dn}(s/n), X_{dn}(t/n))$. We omit the details. We finally use $\mathrm{E}\, X_{dn}(t/n) = n\frac{t^d}{d!}e^{-t} + O(1)$ (uniformly in t).   □

Since $t^d e^{-t}$ has a unique maximum at $t = d$, Theorem 16 now yields the following for the maximal number of vertices of a given degree during the evolution of a random graph.

THEOREM 28. *If $d \geq 1$, then*

$$n^{-1/2}(\max_t X_{dn}(t) - X_{dn}(d/n)) \xrightarrow{p} 0 \tag{11.21}$$

and

$$n^{-1/2}(\max_t X_{dn}(t) - n\frac{d^d}{d!}e^{-d}) \xrightarrow{d} N(0, \frac{d^d}{d!}e^{-d}(1 - \frac{d^d}{d!}e^{-d})).   □$$

We have so far considered a fixed degree d, but we can also consider the joint distribution for several, or all, d. Joint convergence of infinitely many variables will be interpreted as convergence of finite-dimensional distributions below.

REMARK 11.3. We thus consider convergence in $\mathbf{R}^\infty$. This may be improved to convergence in stronger topologies, for example (weighted) l^1, which allows for more continuous functionals, but we leave such extensions to the reader.

In the theorems below we use the notation

$$\pi_i(\lambda) = \frac{\lambda^i}{i!} e^{-\lambda}$$

for the Poisson probabilities, and let $\alpha_{dn} = \mathrm{E}\,\psi_d(G_{n,p_n})$ (which may be replaced by $n\pi_d(np_n)$ in the results). We begin with the case $np_n \to 0$.

THEOREM 29.

(i) If $np_n \to 0$, then (11.12) holds jointly for any set of $d \geq 0$ such that $n(np_n)^{d\vee 1} \to \infty$, with limits $Z_d \sim N(0,1)$ such that $Z_0 = -Z_1$ while $\{Z_d : d \geq 1\}$ are independent.

(ii) If $m_n/n \to 0$, then (11.15) holds jointly for any set of $d \geq 0$ such that $n(m_n/n)^{d\vee 2} \to \infty$, with limits $Z'_d \sim N(0,1)$ such that $Z'_0 = -Z'_1 = Z'_2$, while $\{Z'_d : d \geq 2\}$ are independent.

Proof. For (i) we may use a multi-dimensional version of Theorem 2, see Remark 5.1, noting that $X_{dn}(p_n)$ is dominated by $\{K_{1,d}\}$ for $d \geq 1$, and that $U(K_{1,d};1)$, $d \geq 1$, are independent by Theorem 1. Further, $X_{0n}(p_n)$ is dominated by $\{K_{1,1}\}$, so both Z_0 and Z_1 may be written as constants times $U(K_{1,1};1)$, whence $Z_0 = \pm Z_1$. An inspection of (11.7) shows that actually $Z_0 = -Z_1$.

For (ii) we argue similarly, using a multi-dimensional version of Theorem 9(ii), or Remark 6.4 which leads to $Z_d = c_d U(K_{1,d\vee2};1)$.   □

REMARK 11.4. The normalizations in Theorem 29 depend on d. The results for small d may also be formulated as

$$(\psi_d(G_{n,p_n}) - \alpha_{dn})/(n^2 p_n)^{1/2} \to \begin{cases} -U(K_2;1), & d = 0, \\ U(K_2;1), & d = 1, \\ 0, & d \geq 2, \end{cases}$$

and, with $p_n = m_n/\binom{n}{2}$,

$$(\psi_d(G_{n,m_n}) - \alpha_{dn})/(n^3 p_n^2)^{1/2} \to \begin{cases} \frac{1}{2}U(P_2;1), & d = 0, \\ -U(P_2;1), & d = 1, \\ \frac{1}{2}U(P_2;1), & d = 2. \end{cases}$$

We leave to the reader to see how the coefficients here reflect the fact that the total number of vertices $\sum_d \psi_d$ is fixed, and that for $G_{n,m}$ also the number of edges $\frac{1}{2}\sum_d d\psi_d$ is fixed.

The next result is also a generalization of parts of Theorems 25 and 26.

THEOREM 30.

(i) *Suppose that* $np_n \to \lambda$, $0 < \lambda < \infty$. *Then*

$$n^{-1/2}(\psi_j(G_{n,p_n}) - \alpha_{jn}) \xrightarrow{d} Z_j,$$

jointly in $j \geq 0$, *where* Z_j *are jointly normal,* $\mathrm{E}\, Z_j = 0$ *and*

$$\mathrm{Cov}(Z_i, Z_j) = \delta_{ij}\pi_i(\lambda) - \pi_i(\lambda)\pi_j(\lambda) + \frac{(i-\lambda)(j-\lambda)}{\lambda}\pi_i(\lambda)\pi_j(\lambda). \qquad (11.22)$$

(ii) *Suppose that* $2m_n/n \to \lambda$, $0 < \lambda < \infty$, *and let* $p_n = m_n/\binom{n}{2}$. *Then*

$$n^{-1/2}(\psi_j(G_{n,m_m}) - \alpha_{jn}) \xrightarrow{d} Z'_j,$$

jointly in $j \geq 0$, *where* Z'_j *are jointly normal,* $\mathrm{E}\, Z'_j = 0$ *and*

$$\mathrm{Cov}(Z'_i, Z'_j) = \delta_{ij}\pi_i(\lambda) - \pi_i(\lambda)\pi_j(\lambda) - \frac{(i-\lambda)(j-\lambda)}{\lambda}\pi_i(\lambda)\pi_j(\lambda). \qquad (11.23)$$

Proof. (i) The joint convergence follow by Theorem 2, with

$$Z_j = \sum_{k=1}^{\infty}(-1)^{j+k}\pi_j(\lambda)\frac{\lambda^{k/2}}{k!}C_j(\lambda;k)U(K_{1,k};1), \qquad (11.24)$$

cf. (11.20). Since $\mathrm{aut}(K_{1,k}) = k!$ for $k \geq 2$, but $\mathrm{aut}(K_{1,1}) = 2$,

$$\mathrm{Cov}(Z_i, Z_j) = \sum_{k=1}^{\infty}(-1)^{i+j}\pi_i(\lambda)\pi_j(\lambda)\frac{\lambda^k}{k!}C_i(\lambda;k)C_j(\lambda;k)(1+\delta_{1k})$$

$$= (-1)^{i+j}\pi_i(\lambda)\pi_j(\lambda)\Big(e^{\lambda}\sum_{k=0}^{\infty}C_i(\lambda;k)C_j(\lambda;k)\pi_k(\lambda)$$

$$\qquad - C_i(\lambda;0)C_j(\lambda;0) + \lambda C_i(\lambda;1)C_j(\lambda;1)\Big), \qquad (11.25)$$

and (11.22) follows using the orthogonality relation (11.9).

(ii) Similar, using Theorem 9(iii), which gives Z'_j as in (11.24), but summing only over $k \geq 2$.   □

REMARK 11.5. The covariance (11.22) may alternatively be obtained as the limit of $\mathrm{Cov}(X_{in}(p_n), X_{jn}(p_n))/n$. The calculation above then proves (11.9), which shows that the polynomials in (11.8) have to be orthogonal for $\mathrm{Po}(\lambda)$.

REMARK 11.6. It follows easily from (11.22) that $\sum_{j=0}^{\infty} Z_j = 0$, which reflects the fact that the total number of vertices is fixed. Similarly, $\sum_{j=0}^{\infty} Z'_j = \sum_{j=0}^{\infty} jZ'_j = 0$, as expected because both the number of vertices and the number of edges are fixed in $G_{n,m}$. In fact, the distribution of $(Z'_j)_0^{\infty}$ equals the conditional distribution of $(W_j)_0^{\infty}$ given $\sum_0^{\infty} W_j = \sum_0^{\infty} jW_j = 0$, where $W_j \sim N(0, \pi_j(\lambda))$ are independent.

The proof of Theorem 27 immediately yields joint convergence too. A calculation of the covariances yields the following result, which also includes Theorem 30(i).

THEOREM 31. *The convergence in Theorem 27 holds jointly for all $d \geq 0$. $Z_d(t)$ are jointly Gaussian, with*

$$\mathrm{Cov}\big(Z_i(s), Z_j(t)\big) = \pi_i(s)\pi_{j-i}(t-s) - \pi_i(s)\pi_j(t) + \frac{(s-i)(t-j)}{t}\pi_i(s)\pi_j(t),$$

for $0 \leq s \leq t$ (where $\pi_l(\lambda) = 0$ for $l < 0$).   □

As an illustration of the possibility to consider different times simultaneously, we finally give the multi-dimensional version of Theorem 28.

THEOREM 32.
$$n^{-1/2}\big(\max_t X_{jn}(t) - n\pi_j(j)\big) \xrightarrow{d} W_j,$$

jointly for $j \geq 1$, where W_j are normal with $\mathrm{E}\,W_j = 0$ and

$$\mathrm{Cov}(W_i, W_j) = \pi_i(i)\big(\pi_{j-i}(j-i) - \pi_j(j)\big), \qquad 1 \leq i \leq j.$$

Proof. By (11.21), we may instead consider $n^{-1/2}\big(X_{jn}(j/n) - n\pi_j(j)\big)$, and the result follows from Theorem 31, letting $W_j = Z_j(j)$.   □

12. Further examples

EXAMPLE 12.1. Let $\psi(G)$ be the number of isolated edges in G, and $X_n(t) = \psi(G_n(t))$. It is easily seen that if, for example, $nt \to \lambda > 0$,

$$\mathrm{E}\,X_n(t) = \binom{n}{2}t(1-t)^{2n-4} \sim \tfrac{1}{2}n\lambda e^{-2\lambda} \tag{12.1}$$

and

$$\mathrm{Var}\,X_n(t) \sim n\big(\tfrac{1}{2}\lambda e^{-2\lambda} + (\lambda^3 - \lambda^2)e^{-4\lambda}\big), \tag{12.2}$$

with similar (but simpler) expressions holding for $nt \to 0$ and $nt \to \infty$, $nt^2 \to 0$.

An argument similar to the one in Section 11 gives the coefficients $\widehat{\psi}_n(H; t)$; the support of ψ consists of the empty graph, all stars, and all graphs without isolated vertices that consist of two central vertices, possibly joined by an edge, together with any number of further vertices joined to at least one of the central ones. Some of these graphs are disconnected (the disjoint unions of two stars), but if we restrict ourselves to the case $np_n = O(1)$, it may easily be shown using Proposition 4.7, or Propositions 4.10 and 4.6, that ψ is almost finitely dominated by the family of stars and 'doublestars' (two stars connected by an edge between the centres). If $np_n \to 0$, ψ is dominated by $\{K_2\}$.

The drift equals $\xi(t) = (1-t)^{-1}\big(\tfrac{1}{2}\psi_0(\psi_0 - 1) - (2n-4)\psi\big)(G_n(t))$, where ψ_0 is the number of isolated vertices, and $\mathrm{Var}\,\xi(t) = O\big(n^3(1 + (nt)^3)e^{-2nt}\big)$ for $n \geq 3$. As in the previous section we obtain limit theorems with normal limits for G_{n,p_n} with $np_n \to \lambda \geq 0$ and $n^2 p_n \to \infty$ (a more general result is proved by Barbour, Karoński and Ruciński [3]), and for G_{n,m_n} with $m_n/n \to \lambda \geq 0$ and $m_n^2/n \to \infty$; a functional limit theorem; and a limit theorem for the maximum. We state only the latter, which follows by Theorems 14(ii) and 16, and answers a question by Paul Erdős (personal communication).

THEOREM 33. *Let Y_n be the maximum number of isolated edges during the evolution of a random graph. Then*

$$n^{1/2}(Y_n - \tfrac{1}{4e}n) \xrightarrow{d} N(0, \tfrac{1}{4e} - \tfrac{1}{8e^2}). \qquad \square$$

Similar arguments ought to work for the number of isolated copies of any given tree, cf. [3] where asymptotic normality is proved using Stein's method, but the decomposition becomes somewhat more complicated and we have not checked the details.

EXAMPLE 12.2. Let $\psi(G)$ be the number of spanning subtrees of G. It is not difficult to see, arguing as in Section 10 or using (4.4), that the support of ψ is the set of all forests (without isolated vertices) and that if F is a forest with $v(F) \leq n$ and components of sizes $v_1, ..., v_r \geq 2$, then, because F is contained in $n^{n-2} \prod v_i n^{1-v_i}$ spanning subtrees of the complete graph K_n, see Moon [21],

$$\widehat{\psi}_n(F; p) = \frac{1}{\text{aut}(F)} n^{n-2} \prod_1^r (v_i n^{1-v_i}) p^{n-1-e(F)}$$

$$= \frac{1}{\text{aut}(F)} n^{n-2} p^{n-1} \prod_1^r v_i (np)^{1-v_i}. \qquad (12.3)$$

In particular, with $\text{E}\,\psi = \text{E}\,\psi(G_n(p)) = n^{n-2} p^{n-1}$,

$$\hat{\psi}_n^*(F; p) \asymp \text{E}\,\psi \prod_1^r n^{1-v_i/2} p^{(1-v_i)/2}. \qquad (12.4)$$

Let $p \in (0, 1)$ be fixed. Then (12.4) shows that $\hat{\psi}_n^*(F; p)$ is of order $\text{E}\,\psi$ if all $v_i = 2$, but smaller otherwise. We apply Proposition 4.7 with $\mathcal{H} = \{kK_2 : k \geq 1\}$ and $\beta_n = \text{E}\,\psi$, and obtain $a(kK_2) = \frac{1}{2^k k!} 2^k p^{-k/2} = (k!)^{-1} p^{-k/2}$ and

$$\sum_{\mathcal{H}} a(H)^2 \, \text{aut}(H)(1-p)^{e(H)} = \sum_{k=1}^{\infty} \frac{2^k}{k!} p^{-k}(1-p)^k = e^{2(1-p)/p} - 1.$$

Since $\text{Var}\,\psi(G_{n,p})/(\text{E}\,\psi)^2$ converges to this, see e.g. [9], ψ is almost finitely dominated by $\mathcal{H}$. Theorems 2 and 1 yield, using a well-known generating function for the Hermite polynomials,

$$\psi(G_{n,p})/\text{E}\,\psi \xrightarrow{d} \sum_{k=0}^{\infty} \frac{1}{k!} p^{-k/2} U(kK_2; 1)$$

$$= \sum_{k=0}^{\infty} \frac{1}{k!} p^{-k/2} {:} U(K_2; 1)^k {:}$$

$$= \exp\left(p^{-1/2} U(K_2; 1) - \tfrac{1}{2} p^{-1} \text{Var}\, U(K_2; 1)\right). \qquad (12.5)$$

Since $U(K_2; 1) \sim N(0, 2(1-p))$, this yields, with $q = 1 - p$,

$$\log(\psi(G_{n,p})/\text{E}\,\psi) \xrightarrow{d} N(-\tfrac{q}{p}, \tfrac{2q}{p}). \qquad (12.6)$$

In other words, $\psi(G_{n,p})$ has asymptotically a log-normal distribution. (This is proved by other means in [9].)

Similar results hold for the number of Hamiltonian cycles and the number of perfect matchings, cf. [12].

The theorems above are not immediately applicable when $p \to 0$, since ψ then is not almost finitely dominated. In [12] we show how a modification of ψ gives results for this case, and also for $G_{n,m}$. (The modification is similar to the ones in Section 6, but now we *multiply* $\psi(G)$ by a suitable function of $e(G)$.)

EXAMPLE 12.3. Let $\phi(G) = \mathrm{sub}(K_3, G)$ and let $\chi(G) = \binom{n}{3} \frac{(e(G))_3}{(\binom{n}{2})_3}$. Thus $\chi(G_{n,m}) = \mathrm{E}\,\phi(G_{n,m})$, and if we define $\psi = \phi - \chi$, $\mathrm{E}\,\psi(G_{n,m}) = 0$ for every m. Thus also $\mathrm{E}\,\psi(G_{n,p}) = 0$ for every p.

By Proposition 4.2, $\widehat{\psi}_n(\emptyset; t) = \widehat{\psi}_n(K_2; t) = 0$. Furthermore, by arguments as above, $\widehat{\psi}_n(P_2; t) = \frac{1}{2}t + O(n^{-1})$; $\widehat{\psi}_n(K_3; t) = \frac{1}{6} + O(n^{-3})$; $\widehat{\psi}_n(K_2 + K_2; t) = O(n^{-1})$; $\widehat{\psi}_n(H; t) = O(n^{-3})$ for $H \in \{P_3, K_{1,2}, P_2 + K_2, 3K_2\}$; and $\widehat{\psi}_n(H; t) = 0$ otherwise. Theorem 12 applies with $p_n = 1$ and $\beta_n = n^{3/2}$ and yields

$$n^{-3/2}\psi(G_n(t)) \xrightarrow{d} Z(t) \qquad \text{in } D([0, \infty]), \tag{12.7}$$

with

$$Z(t) = \tfrac{t}{2}U(P_2; t) + \tfrac{1}{6}U(K_3; t). \tag{12.8}$$

Consequently, $Z(t)$ is a continuous Gaussian process with

$$\begin{aligned} \mathrm{Cov}(Z(s), Z(t)) &= \tfrac{1}{2}st(s(1-t))^2 + \tfrac{1}{6}(s(1-t))^3 \\ &= \tfrac{1}{6}s^3(1+2t)(1-t)^2, \qquad 0 \le s \le t \le 1. \end{aligned} \tag{12.9}$$

It follows immediately that

$$n^{-3/2}\max_t \psi(G_n(t)) \xrightarrow{d} \max_{0 \le t \le 1} Z(t). \tag{12.10}$$

Note that the maximum in this case behaves quite differently from the cases covered by Theorems 16 and 17.

We finally refer to [2] for yet another example.

REFERENCES

1. A. D. Barbour, L. Holst and S. Janson, *Poisson Approximation*, Oxford Univ. Press, Oxford, 1992.
2. A. D. Barbour, S. Janson, M. Karoński and A. Ruciński, *Small cliques in random graphs*, Random Struct. Alg. **1** (1990), 403–434.
3. A. D. Barbour, M. Karoński and A. Ruciński, *A central limit theorem for decomposable random variables with applications to random graphs*, J. Combin. Th. Ser. B **47** (1989), 125–145.
4. P. Billingsley, *Convergence of Probability Measures*, Wiley, New York, 1968.
5. B. Bollobás, *Random Graphs*, Academic Press, London, 1985.
6. P. Erdős and A. Rényi, *On the evolution of random graphs*, Magyar Tud. Akad. Mat. Kut. Int. Közl. **5** (1960), 17–61.
7. S. N. Ethier and T. G. Kurtz, *Markov processes*, Wiley, New York, 1986.
8. J. Jacod and A. N. Shiryaev, *Limit Theorems for Stochastic Processes*, Springer, Berlin, 1987.
9. S. Janson, *Random trees in a graph and trees in a random graph*, Math. Proc. Camb. Phil. Soc. **100** (1986), 319–330.
10. S. Janson, *A functional limit theorem for random graphs with applications to subgraph count statistics*, Random Struct. Alg. **1** (1990), 15–37.
11. S. Janson, *A graph Fourier transform and proportional graphs*, Random Graphs '91, special issue of Random Struct. Alg. (to appear).
12. S. Janson, *The numbers of spanning trees, Hamilton cycles and perfect matchings in a random graph* (to appear).
13. S. Janson and J. Kratochvíl, *Proportional graphs*, Random Struct. Alg. **2** (1991), 209–224.
14. S. Janson, T. Łuczak and A. Ruciński, *An exponential bound for the probability of nonexistence of a specified subgraph in a random graph*, Random Graphs '87, Wiley, Chichester, 1990, pp. 73–87.
15. S. Janson and K. Nowicki, *The asymptotic distributions of generalized U-statistics with applications to random graphs*, Probab. Th. Rel. Fields **90** (1991), 341–375.
16. S. Janson and J. Spencer, *Probabilistic construction of proportional graphs*, Random Struct. Alg. **3** (1992), 127–137.
17. M. Karoński and A. Ruciński, *Poisson convergence and semi-induced properties of random graphs*, Math. Proc. Camb. Phil. Soc. **101** (1987), 291–300.
18. J. Kärrman, *Existence of proportional graphs*, J. Graph Th. **17** (1993), 207–220.
19. J. Kärrman, *An example of a superproportional graph*, Random Graphs '91, special issue of Random Struct. Alg. (to appear).

20. T. Lindvall, *Weak convergence of probability measures and random functions in the function space* $D[0, \infty)$, J. Appl. Prob. **10** (1973), 109–121.

21. J. W. Moon, *Enumerating labelled trees*, Graph theory and theoretical physics, ed. F. Harary, Academic Press, London, 1967, pp. 261–272.

22. B. Pittel, *On tree census and the giant component in sparse random graphs*, Random Struct. Alg. **1** (1990), 311–342.

23. P. Protter, *Stochastic integration and differential equations*, Springer, Berlin, 1990.

24. A. Ruciński, *When are small subgraphs of a random graph normally distributed?*, Prob. Th. Rel. Fields. **78** (1988), 1–10.

25. V. E. Stepanov, *On the probability of connectedness of a random graph* $\mathcal{G}_m(t)$, Teor. Veroyatnost. i Primenen. **15** (1970), 55–67, (Russian); Theory Probab. Appl. **15** (1970), 55–67.

DEPARTMENT OF MATHEMATICS, UPPSALA UNIVERSITY, PO BOX 480, S-751 06 UPP-SALA, SWEDEN

E-mail address: svante.janson@math.uu.se

Editorial Information

To be published in the *Memoirs*, a paper must be correct, new, nontrivial, and significant. Further, it must be well written and of interest to a substantial number of mathematicians. Piecemeal results, such as an inconclusive step toward an unproved major theorem or a minor variation on a known result, are in general not acceptable for publication. *Transactions* Editors shall solicit and encourage publication of worthy papers. Papers appearing in *Memoirs* are generally longer than those appearing in *Transactions* with which it shares an editorial committee.

As of June 7, 1994, the backlog for this journal was approximately 7 volumes. This estimate is the result of dividing the number of manuscripts for this journal in the Providence office that have not yet gone to the printer on the above date by the average number of monographs per volume over the previous twelve months, reduced by the number of issues published in four months (the time necessary for preparing an issue for the printer). (There are 6 volumes per year, each containing at least 4 numbers.)

A Copyright Transfer Agreement is required before a paper will be published in this journal. By submitting a paper to this journal, authors certify that the manuscript has not been submitted to nor is it under consideration for publication by another journal, conference proceedings, or similar publication.

Information for Authors and Editors

Memoirs are printed by photo-offset from camera copy fully prepared by the author. This means that the finished book will look exactly like the copy submitted.

The paper must contain a *descriptive title* and an *abstract* that summarizes the article in language suitable for workers in the general field (algebra, analysis, etc.). The *descriptive title* should be short, but informative; useless or vague phrases such as "some remarks about" or "concerning" should be avoided. The *abstract* should be at least one complete sentence, and at most 300 words. Included with the footnotes to the paper, there should be the 1991 *Mathematics Subject Classification* representing the primary and secondary subjects of the article. This may be followed by a list of *key words and phrases* describing the subject matter of the article and taken from it. A list of the numbers may be found in the annual index of *Mathematical Reviews*, published with the December issue starting in 1990, as well as from the electronic service e-MATH [**telnet e-MATH.ams.org** (or **telnet 130.44.1.100**). Login and password are **e-math**]. For journal abbreviations used in bibliographies, see the list of serials in the latest *Mathematical Reviews* annual index. When the manuscript is submitted, authors should supply the editor with electronic addresses if available. These will be printed after the postal address at the end of each article.

Electronically prepared manuscripts. The AMS encourages submission of electronically prepared manuscripts in $\mathcal{A}_{\mathcal{M}}\mathcal{S}$-TEX or $\mathcal{A}_{\mathcal{M}}\mathcal{S}$-LATEX because properly prepared electronic manuscripts save the author proofreading time and move more quickly through the production process. To this end, the Society has prepared "preprint" style files, specifically the amsppt style of $\mathcal{A}_{\mathcal{M}}\mathcal{S}$-TEX and the amsart style of $\mathcal{A}_{\mathcal{M}}\mathcal{S}$-LATEX, which will simplify the work of authors and of the production staff. Those authors who make use of these style files from the beginning of the writing process will further reduce their own effort. Electronically submitted manuscripts prepared in plain T_EX or LaT_EX do not mesh properly with the AMS production systems and cannot, therefore, realize the same kind of expedited processing. Users of plain T_EX should have little difficulty learning $\mathcal{A}_{\mathcal{M}}S$-T_EX, and LaT_EX users will find that $\mathcal{A}_{\mathcal{M}}S$-LaT_EX is the same as LaT_EX with additional commands to simplify the typesetting of mathematics.

Guidelines for Preparing Electronic Manuscripts provides additional assistance and is available for use with either $\mathcal{A}_{\mathcal{M}}S$-T_EX or $\mathcal{A}_{\mathcal{M}}S$-LaT_EX. Authors with FTP access may obtain *Guidelines* from the Society's Internet node e-MATH.ams.org (130.44.1.100). For those without FTP access *Guidelines* can be obtained free of charge from the e-mail address guide-elec@math.ams.org (Internet) or from the Customer Services Department, American Mathematical Society, P.O. Box 6248, Providence, RI 02940-6248. When requesting *Guidelines*, please specify which version you want.

At the time of submission, authors should indicate if the paper has been prepared using $\mathcal{A}_{\mathcal{M}}S$-T_EX or $\mathcal{A}_{\mathcal{M}}S$-LaT_EX. The *Manual for Authors of Mathematical Papers* should be consulted for symbols and style conventions. The *Manual* may be obtained free of charge from the e-mail address cust-serv@math.ams.org or from the Customer Services Department, American Mathematical Society, P.O. Box 6248, Providence, RI 02940-6248. The Providence office should be supplied with a manuscript that corresponds to the electronic file being submitted.

Electronic manuscripts should be sent to the Providence office immediately after the paper has been accepted for publication. They can be sent via e-mail to pub-submit@math.ams.org (Internet) or on diskettes to the Publications Department, American Mathematical Society, P.O. Box 6248, Providence, RI 02940-6248. When submitting electronic manuscripts please be sure to include a message indicating in which publication the paper has been accepted.

Two copies of the paper should be sent directly to the appropriate Editor and the author should keep one copy. The *Guide for Authors of Memoirs* gives detailed information on preparing papers for *Memoirs* and may be obtained free of charge from the Editorial Department, American Mathematical Society, P.O. Box 6248, Providence, RI 02940-6248. For papers not prepared electronically, model paper may also be obtained free of charge from the Editorial Department.

Any inquiries concerning a paper that has been accepted for publication should be sent directly to the Editorial Department, American Mathematical Society, P.O. Box 6248, Providence, RI 02940-6248.

Recent Titles in This Series

(*Continued from the front of this publication*)

(See the AMS catalog for earlier titles)